Transcendence and Spatiality of th

European University Studies
Europäische Hochschulschriften
Publications Universitaires Européennes

Series XXIII
Theology

Reihe XXIII   Série XXIII
Theologie
Théologie

Vol./Band 815

PETER LANG
Bern · Berlin · Bruxelles · Frankfurt am Main · New York · Oxford · Wien

Chan Ho Park

Transcendence and Spatiality of the Triune Creator

PETER LANG

Bern · Berlin · Bruxelles · Frankfurt am Main · New York · Oxford · Wien

Bibliographic information published by Die Deutsche Bibliothek
Die Deutsche Bibliothek lists this publication in the Deutsche Nationalbibliografie;
detailed bibliographic data is available on the Internet at
‹http://dnb.ddb.de›.

British Library and Library of Congress Cataloguing-in-Publication Data:
A catalogue record for this book is available from *The British Library*, Great Britain,
and from *The Library of Congress*, USA

ISSN 0721-3409
ISBN 3-03910-844-1

© Peter Lang AG, European Academic Publishers, Bern 2005
Hochfeldstrasse 32, Postfach 746, CH-3000 Bern 9, Switzerland
info@peterlang.com, www.peterlang.com, www.peterlang.net

All rights reserved.
All parts of this publication are protected by copyright.
Any utilisation outside the strict limits of the copyright law, without
the permission of the publisher, is forbidden and liable to prosecution.
This applies in particular to reproductions, translations, microfilming,
and storage and processing in electronic retrieval systems.

Printed in Germany

Acknowledgments

This book was originally presented to Fuller Theological Seminary as my Ph.D dissertation in 2003. Special appreciation is given to Dr. William A. Dyrness to whom I owed much insight about this project. He guided and advised me kindly and patiently during the writing of my dissertation. My original theological interest was on panentheism of contemporary theologians. By the help of the book of Dyrness *The Earth is God's*, I could narrow down my focus into God and space. I also want to give my thanks to Dr. Veli-Matti Kärkkäinen and late Dr. Colin E. Gunton. Especially Dr. Gunton had read this book as one of the examiners three months before his death. I dedicate this book to Dr. Gunton.

The material and spiritual supports of Springing Fountain Korean Church in LA, California, and Bundang Central Church in Korea were invaluable. When I stayed in LA district to study in Fuller Seminary, I usually had the privilege to preach a sermon per week. It helped me keep my spiritual strength and gave an incentive to my study. I give thanks to their warm encouragement of Springing Fountain Church. For nearly three years of the final period of my project, Bundang Central Church did not give only the financial support but also prayed for my study. In order to raise fund for supporting the Korean students who are scattered worldwidely, that church sacrificed many parts, especially constructing a new church building. Westminster Graduate School of Theology in Seoul offered me a teaching position even before I took my doctoral degree. Teaching experience in that school is so pleasant to me. I give thanks to its all staffs, professors, and students. By means of their warm acceptance, I could finish this project without being bothered. I give my special thanks to Chairman of Board of Trustee of the school, Cha Saeng Kim. He gladly supported the expense of this book. Since last December I have served as the President of the school.

I dedicate this work to my family. My wife Shin Hwa is patient and supports me throughout this project. We have suffered infertility for more than 12 years. Now in her womb is a baby. What a marvelous grace it is! For seven years my family has given an unchanging support. Of course deepest thanks belong to God. I always remember this verse of

VI   *Acknowledgments*

Psalms: "To all perfection I see a limit; but your commands are boundless" (Ps. 119: 96).

August 31, 2005

Table of Contents

Chapter I
Introduction 1
1. Classical Theism and Panenthesim.. 1
2. Divine Transcendence and Spatiality.. 5
3. Thesis and Plan... 14

Chapter II
Concepts of Space.. 19
1. Introduction 19
2. Two Concepts of Space 21
3. Controversy over Absolute Space... 23
4. Concept of Space in the Theory of Relativity.................................. 26
5 A Reformed Perspective on Space.. 32
6 Perichoretic Concept of Space.. 39
7. Conclusion... 42

Chapter III
Divine Transcendence.. 45
1. Introduction 45
2. Crisis of Transcendence.. 46
3. Transcendence versus Panentheism.. 52
4. Immanence versus Deism... 61
5 Transcendence and Spatiality 66
6. Conclusion... 74

Chapter IV
God, Space and Spatiality: Case Studies.. 77
1. Introduction 77
2. Torrance-Redemption of Space 81
    2.1 God and the World 81
    2.2 Redemption of Space and Time 85

VIII
Table of Contents

2.3 God's Temporality	90
2.4 Critical Evaluation	92
3. Pannenberg – Spirit as a Force Field	95
3.1 Spirit as a Force Field	96
3.2 Time and Space	99
3.3 Omnipresence	102
3.4 Critical Evaluation	103
4. Moltmann-Creation as Divine *Zimzum*	108
4.1 Self-determination of God	108
4.2 God's Self-limitation	113
4.3 The Cosmic *Shekinah*	115
4.4 Critical Evaluation	118
5. Conclusion	124

Chapter V
Divine Spatiality 127

1. Introduction	127
2. God Is Space	129
3. God Has Space	132
4. The Body of God	137
5. 'The Earth Is God's'	147
6. Divine Action and Embodiment	151
7. God Is in Space	159
8. Conclusion	164

Chapter VI
The Presence of the Triune God 167

1. Introduction	167
2. Relationship	170
2.1 The Renaissance of Trinitarian Theology	171
2.2 Theology of Relationship	175
2.3 The Relation of God to the Creation	183
3. Agency	195
3.1 Interventionism	196
3.2 Top-Down Causation	202
3.3 Bottom-Up Causation Revisited	206
4. Embodiment	221
4.1 Creation	222

Table of Contents IX

    4.2 Incarnation 227
    4.3 The Church 233
    4.4 Eucharist 240
    4.5 New Creation 250
5. Conclusion 258

Conclusion
Summary and Case for Classical Theism 261

Bibliography 269

Chapter I

Introduction

1. Classical Theism and Panentheism

One special characteristic of contemporary theology is its critical reflection on the traditional concepts of God. Thus, in the writings of the contemporary theologians we find arguments concerning the attributes of God, such as God's eternity and God's immutability. According to the reflection of contemporary theologians, some elements of the traditional concepts of God are influenced by philosophical concepts at odds with biblical tradition. In the formulation of Christian theology, there was a widespread influence of Greek philosophy. As a result, theologians argue, the concepts of God in traditional or classical theology often do not fit the biblical narrative. For example, the God of the Bible has interacted in the events of this world within space and time. God is not only transcendent beyond the creation but also immanent in world history. For contemporary theologians, traditional theology cannot reckon with these activities of the God of the Bible. Thus, many theologians have tried to revise the traditional concepts of God.

Process theology has become one model for the efforts of contemporary theologians. Since many scholars regard process theology as an alternative to traditional theology, it has been labeled "neo-classical."[1]

---

1 Charles Hartshorne was the first person to call his own panentheistic version of process theology "neoclassical theism." See Ewert H. Cousins, ed., *Process Theology* (New York: Newman Press, 1971), p. 47. Cf. Colin E. Gunton, *Becoming and Being: The Doctrine of God in Charlse Hartshorne and Karl Barth* (London: SCM Press, 2001), p. 24. Ronald H. Nash, *The Concept of God: An Exploration of Contemporary Difficulties with the Attributes of God* (Grand Rapids: Zondervan, 1983), p. 19. Sometimes process theologians identify their orientation as "dipolar theism," in contrast to traditional theism with its doctrine of divine simplicity [John B. Cobb, Jr. and David Ray Griffin, *Process Theology: An Introductory Exposition* (Philadelphia: Westminster Press, 1976), p. 47].

2 *Transcendence and Spatiality of the Triune Creator*

However, process theology asserts some doctrines which are evidently contrary to the Christian faith. For example, it denies the doctrine of creation *ex nihilo*. It also rejects the eschaton of this world. For many theologians these positions are too far from what is orthodox about traditional theology. They have a strong sense of the need to revise the traditional concepts of God, but they have sought a way between classical theism and process theology.[2]

Some theologians refute any attempts to revise some concepts of God of classical theism. Even though Richard A. Muller recognizes the significance of the challenges from revisionists, he wants to "remain convinced that the structure of classical theism and its synthesis of faith and reason stands the test and survives the critique."[3] Along with Muller, John Polkinghorne does not feel the need to withdraw the framework of classical theism despite its undue emphasis on the transcendence of God. In comparison to Muller, however, he admits the need to revise some traditional concepts of God. Polkinghorne requires "a recovery of the balancing orthodox concept of divine immanence."[4] In his book, *The Concept of God*, Ronald H. Nash contrasts two concepts of God: classical theism, Christian theism or Thomistic theism *versus* panentheism, neoclassical theism, or process theology. Both the traditional and the revised concepts have their merits and demerits. But neither is an appropriate option for Christian theology. The alternative between Thomistic theism and panentheism is not a necessary but a forced and artificial choice. Rather, as Nash argues, "It is possible to develop mediating con-

---

2   We see such attempts in the following books by Ted Peters, *God as Trinity: Relationality and Temporality in Divine Life* (Louisville, KY: Westminster/John Knox Press, 1993) and Clark Pinnock, et al., *The Openness of God: A Biblical Challenge to the Traditional Understanding of God* (Downers Grove, IL: InterVarsity Press, 1994). For me, "openness theists," or "the free will theists," who are "impressed with the wisdom implicit in the 'Wesleyan quadrilateral' of Scripture, tradition, reason, and experience" go as far as process theologians in their later phase. See John B. Cobb, Jr. and Clark H. Pinnock, eds, *Searching for an Adequate God: A Dialogue between Process and Free Will Theists* (Grand Rapids: Eerdmans, 2000).

3   Richard A. Muller, "Incarnation, Immutability, and the Case for Classical Theism," in *Westminster Theological Journal* 45 (1983), p. 24. He mainly attacks the article of Clark Pinnock, "The Need for a Scriptural, and Therefore a Neo-Classical Theism," in Kenneth Kantzer and Stanley Gundry, eds, *Perspectives on Evangelical Theology* (Grand Rapids: Baker, 1979), pp. 37–42.

4   John Polkinghorne, *Scientists as Theologians: A Comparison of the Writings of Ian Barbour, Arthur Peacocke and John Polkinghorne* (London: SPCK, 1996), p. 33.

Introduction

3

cepts of God that can avoid the major difficulties of the static God of Thomistic theism and the finite god of Process theology."[5] It is the purpose of this dissertation to follow up the assertions of Polkinghorne and Nash, that is, to grope for the possibility of divine spatiality within the framework of classical theism.

The doctrine of God of process theologians is characterized as panentheism. The term panentheism was made and proposed in a relatively modern time. The German philosopher Karl C. F. Krause (1781–1832) is regarded as the person who introduced the term panentheism. He was a student of Hegel and Fichte. In order to distinguish his own doctrine of God from contemporary forms of pantheism and emanationism, Krause selected the term panentheism. According Charles Hartshorne, however, there seems no evidence that Krause had much influence on later development of the term, apart from furnishing a label.[6]

Process theologians contend that God is dependent upon the world, as this world is dependent upon Him. In opposition to this contention, so-called "classical theism" asserts that God by nature is independent of this world, even though this world of creatures is absolutely dependent upon God the Creator and God is active in relation to the world. The panentheism of process theology denies the doctrine of creation out of nothing. For, "the doctrine of creation means that while the world is dependent upon God, God is not dependent on the world."[7] Thus the relationship between God and the world in the panentheism of process theology is not unilaterally but mutually dependent.

Panentheism holds that God and the world are necessarily correlative: not only does the world need God for its existence; God also needs the world truly to be God. To clarify the nature of panentheism, we differentiate it from both pantheism and classical theism. In contrast to pantheism which maintains that "everything *is* God," the most popular view of

---

5   R. H. Nash, *The Concept of God*, p. 22.

6   Charles Hartshorne, "Pantheism and Panentheism," *The Encyclopedia of Religion*, ed. Mircea Eliade, vol. 11 (New York: Macmillan, 1987), p. 168. For Krause, the deity is a divine organism inclusive of all lesser organisms. Krause said that "God is more than and includes nature and man." "Consciously or not," according to Hartshorne, Krause "was to some extent returning to Plato's *Timaeus*" (ibid.).

7   R. H. Nash, "Process Theology and Classical Theism," in *Process Theology* (Grand Rapids: Baker, 1987), p. 15.

panentheism holds that "everything is *in* God."[8] E. R. Naughton contends that "panentheism (Gr. παν, all; εν, in; θεos, God) views all things as being in God without exhausting the infinity of the divine nature."[9] According to him, panentheism is "a kind of surrelativism holding for a real convertible relation of dependence between God and the world – not only is the world dependent upon God, but He is dependent upon the world."[10]

In order to distinguish panentheism from pantheism, Charles Hartshorne asks this question: "Do the creatures have genuine freedom of decision making, or does God determine everything?" He argues that classical pantheism holds a form of theological determinism. According to Hartshorne, the ancient Greek Stoics and Spinoza are classical pantheists. In contrast to pantheists, panentheists reject determinism. Thus, according to Hartshorne, "panentheists object that, if one power determines all, there is, causally speaking, only one agent in all action."[11]

Hartshorne characterizes panentheism as saying the following about God: "God is not just the all of (other) things; but yet all other things are literally in him... To be himself he does not need *this* universe, but only *a* universe, and only contingently does he even contain this particular actual universe."[12] God needs some world, but not necessarily this one. According to this definition of Hartshorne, therefore, God is logically independent of this world but not ontologically independent of all worlds.

The issue here is God's aseity. As in classical theism, God needs nothing outside himself to be completely and actually God. In contrast to classical theism, however, panentheism distinguishes God and the world, and it affirms that God has an essential nature that does not formally change. But that essential nature is not completely actual. God himself can grow, develop, and change by actualizing the potentiality of the essential divine nature in interaction with his creation. Thus, Naughton

---

8   C. E. Simcox, "Christian Panentheism," *The Christian Challenge*, vol. 32 (April/ May, 1993), p. 21.

9   E. R. Naughton, "Panentheism," *New Catholic Encyclopedia*, vol. X (Washington, D.C.: The Catholic University of America, 1967), p. 943.

10   Naughton, p. 943 f.

11   Hartshorne, "Pantheism and Panentheism," p. 166.

12   C. Hartshorne and W. L. Reese, eds, *Philosophers Speak of God* (Chicago: University of Chicago Press, 1953), p. 22.

Introduction 5

states, "Panentheism is rooted in a conviction that the world as possible in the mind of God becomes actualized and thereby adds to God's actuality. It opposes the Thomistic view of God as Pure Act."[13]

The crucial difference that distinguishes panentheism from classical theism is the dipolarity of God. In classical theism, the qualitative difference between God and the world of creatures is important. While God is eternal, infinite, and immutable, the world is temporal, finite, and mutable. God is a monopolar being. But, for panentheists, God is to be thought of as a dipolar being. God comprises two contrasting poles: the potential pole and the actual pole. "With respect to His potentiality, God is absolute, eternal and infinite. With respect to His actual pole (the world), God is relative, temporal, and finite."[14] God's potential pole is what God is eternally; his actual pole is what God is at any given moment. Panentheists have tried to speak of some divine attributes differently from classical or traditional theists. Naughton connects this dipolar view of God with the dialectics of Hegel: "Panentheists give special importance to what they call a logic of polarity, which has close affinity to Hegelian dialectics, as the only means of escaping ultimate arising from the use of categories."[15]

2. Divine Transcendence and Spatiality

Speaking in traditional language, to say that God is transcendent is to say that God is not a temporal and spatial being; for God is the creator of the world of time and space. The corollary of this thought is to say that God is qualitatively different from the world. On the one hand, this theologi-

---

13 Naughton, p. 944.
14 Nash, *The Concept of God*, p. 28. Cf. Robert R. Williams, *Schleiermacher the Theologian: The Construction of the Doctrine of God* (Philadelphia: Fortress, 1978), p. xi, states that Schleiermacher's concept of God is different from that of classical monopolar theology. It is similar to Nikolaus Cusanus' concept of God, that is, God is a coincidence of opposites. This is a bipolar concept of God. In this sense, Schleiermacher is in a friendly relationship with contemporary process theology.
15 Naughton, p. 944.

6   *Transcendence and Spatiality of the Triune Creator*

cal framework has the merit of easily and clearly explaining the divine attributes, including omniscience, infinity and immutability. Yet, on the other hand, this position is inadequate to speak of God's interaction with the world. If Christian theology does not support a complete dualism between God and the world, then it should have some capacity for commenting on divine action in this world of time and space.

Ray S. Anderson has proposed that the Christian doctrine of divine transcendence should be neither the "out-and-out" kind of transcendence nor the "immanent" kind of transcendence. He labels his version of this theme "historical transcendence." Regardless of whether Anderson's conclusion is defensible, his statement of the problem seems quite accurate: we should avoid the two extreme views of transcendence. Nevertheless, Anderson's historical transcendence is not clearly differentiated from the "immanent" kind of transcendence, which he seeks to avoid.[16] The divine transcendence of panentheism is, in other words, another version of immanent transcendence.

Peter Kreeft and Ronald K. Tacelli comment on one of the features of panentheism as follows: "panentheism is one way of making God temporal." Yet, they argue that it is possible to maintain that "God is somehow in time" and also to say that He is not a panentheistic deity. To defend this position, they propose that we need "some analysis of time that does not involve any lack or incompletion of being."[17] Similar to the way Kreeft and Tacelli argue for the temporality of God, in this book I will argue for a means to regard God as spatial in some sense. Panentheism is definitely one way of making God spatial as well as temporal. But throughout my argument here, I will be suggesting that everyone who would defend an argument for the spatiality of God need not be panentheistic.

William A. Dyrness criticizes theologies of history and narrative, seeking ways to correct and deepen them. He points out that, "History and story are categories that readily incorporate time but not *space*. They imply movement, as in a melody, but they cannot fully comprehend the

---

16   R. S. Anderson, *Historical Transcendence and the Reality of God* (Grand Rapids: Eerdmans, 1975), p. 24. Even though it is debatable whether Anderson is a panentheist, I think he would deny it.

17   P. Kreeft & R. K. Tacelli, *Handbook of Christian Apologetics: Hundreds of Answers to Crucial Questions* (Downers Grove: InterVarsity Press, 1994), p. 94.

Introduction 7

soil – the complexity in which this movement is rooted."[18] Concerning the importance of discussing space, Dyrness contends as follows:

> It is in space that we encounter the other person, with all of the strangeness that not only does not fit into our story but may challenge it in fundamental ways. As John MacMurray points out, our relation to others is primarily tactile, we encounter them but we do not comprehend them. So we must find ways to understand the physical nature of this encounter. For it is in space that our embodied practices are constructed, where we injure or heal the earth, where we shape objects and events that celebrate our deepest values.[19]

Space and time generally denote a limited condition of creation. In this sense, God is absolutely transcendent over the space and time of creatures. On these terms, it is impossible to discuss any relation of God to space and time. God is the Creator of space and time. In order to differentiate God's space from the ordinary meaning of space, I will from here on out use the term "spatiality." Thus, God's spatiality is a concept used here to suggest that He has space other than that of creatures.

This book will deal with the issue of divine spatiality in relation to the issue of divine transcendence. Unlike the temporality of God, this issue of God's spatiality has drawn little attention from scholars. Arthur Peacocke, John Polkinghorne, and Thomas F. Tracy offer specific treatment of the subject of the divine temporality. But none of them refers to divine spatiality. Perhaps this issue of God's spatiality seems too suggestive of his corporeality. Not until recent times have people, including Christian theologians, thought seriously about the body. The traditional assumption has been that God is Spirit, and therefore He cannot have a bodily existence. In Christian theology, as a result, little room has been left for any discussion of God's spatiality. Only three contemporary theologians have offered discussions of any real length on the problem of space in Christian theology. They are Thomas F. Torrance,[20] Jürgen Moltmann[21]

---

18 William A. Dyrness, *The Earth is God's: A Theology of American Culture* (Maryknoll, NY: Orbis Books, 1997), p. 12.
19 Dyrness, *The Earth is God's*, p. 13.
20 Thomas F. Torrance, *Space, Time and Incarnation* (Edinburgh: T&T Clark, 1969).
21 Jürgen Moltmann, *God in Creation* (Minneapolis: Fortress Press, 1993), pp. 140–157, *The Coming of God* (Minneapolis: Fortress Press, 1996), pp. 296–308 and *Experiences in Theology* (Minneapolis: Fortress Press, 2000), pp. 309–323.

8   *Transcendence and Spatiality of the Triune Creator*

and Wolfhart Pannenberg.[22] As another reference, which is related (although indirectly) to Christian theology, we have a book by Max Jammer which introduces the history of theories of space in physics.[23] These sources will be referenced throughout this book.

Colin Gunton opposes what he sees as the temporalistic theists' attempt to read our time into God.[24] At the same time, he upholds a positive relation of God to time. Gunton presents his view as an apophatic one[25]: "we should accept neither the timelessness nor the temporality of the being of God."[26] He argues that we should stop our discussion at this apophatic conclusion. We should not go beyond this point, Gunton warns, for to do so is mere speculation. Although he does not present his view on God's spatiality, it seems to be that Gunton would hold a position similar to that of divine temporality, that is, an apophatic one: we

---

22   Wolfhart Pannenberg, *Toward a Theology of Nature: Essays on Science and Faith* (Louisville, KY: Westminster/John Knox Press, 1993), pp. 15–71 and *Systematic Theology*, vol. 2 (Grand Rapids: Eerdmans, 1994), pp. 76–115.

23   Max Jammer, *Concepts of Space: The History of Theories of Space in Physics* (New York: Dover Publications, Inc., 1993, Third, Enlarged Edition).

24   Colin E. Gunton, *The Triune Creator: A Historical and Systematic Study* (Grand Rapids: Eerdmans, 1998), p. 92. Cf. Nash, *The Concept of God*, p. 73f. Contrary to the view of the classical theists, such as Augustine, Anselm and Aquinas, that "God's eternal existence is to be understood as timelessness," temporalist theists assert that God's eternal existence is to be interpreted as everlastingness rather than timelessness. Nash names these temporalist theists: Duns Scotus and William of Occam who challenged the atemporalist view; Samuel Clark and Jonathan Edwards who proposed that God is everlasting rather than timeless; Nelson Pike, Nicholas Wolterstorff, Richard Swinburne, William Rowe, and Anthony Kenny as contemporary proponents of the temporalist position.

25   Concerning the term of "apopatic," David F. Ford, ed., *The Modern Theologians: An Introduction to Christian Theology in the Twentieth Century* (Oxford: Blackwell Publishers, 1997), says: "Relating to what is beyond expression in speech; more specifically, relating to the method of negative theology which stresses the transcendence of God over all human language and categories, and prefers forms of reference which say what God is not" (p. 734). T. F. Torrance, *The Christian Doctrine of God: One Being Three Persons* (Edinburgh: T&T Clark, 2001), contrary to Gunton, opposes an apophatic position: "This does not mean that when we reach this point, at the threshold of the Trinity *ad intra*, theological activity simply breaks off, perhaps in favour of some kind of merely apophatic or negative contemplation. While the Triune God is certainly more to be adored than expressed, the ultimate Rationality, as well as the sheer Majesty of God's self-revelation, and above all the Love of God, will not allow us to desist" (p. 111).

26   C. E. Gunton, *The Triune Creator*, p. 92.

Introduction 9

should accept neither the spacelessness nor the spatiality of the being of God. The present study would agree with this statement but I will also attempt to go beyond this apophatic position. Moltmann, Hans Schwarz, Donald Bloesch and Gerald Bray share the assertion that God has his own space. But they differ in the details of explaining such a statement. Since Paul Helm argues for the timeless God, he opposes a view that sees God as in space. God is spaceless as well as timeless. Peter Kreeft and Tacelli also seem to share this view.

It is beyond the scope of this book to offer a detailed discussion of the relation between time and eternity. Yet, in the course of my argument, I will draw from these conversations. In spite of the obvious differences between space and time, the arguments on God's temporality are helpful, by analogy, for structuring a position on God's spatiality. For, unlike the temporality of God, there are few data on the spatiality of God. In his book, *The Earth is God's*, William Dyrness presents three categories that can be used to express the triune presence of God: relationship, agency, and embodiment.[27] I will use these categories to set forth a positive view of the spatiality of God.

Feminist theologians raise some debate on the theological language about God. For feminist theologians including Elizabeth Johnson, the question "what is the right way to speak about God?" is a question of unsurpassed importance. God is holy mystery. The reality of God is beyond all imagining. Johnson contends that since God is an inexhaustible mystery, talk about God must be considered as always being open-ended. For her, new attempts to articulate the concept of the word "God" are to be expected and even welcomed in any and all historical circumstances. She appeals to the authority of Rahner: "it actually postulates thereby a history of our own concept of God that can never be concluded."[28] Through this process, Johnson argues for the right to talk about God from a feminist perspective.

Similar to Johnson, Sallie McFague writes as follows: "since no language about God is adequate and all of it is improper, new metaphors are

---

27 William A. Dyrness, *The Earth is God's: A Theology of American Culture* (Maryknoll, NY: Orbis Books, 1997).

28 Elizabeth A. Johnson, *SHE WHO IS: The Mystery of God in Feminist Theological Discourse* (New York: Crossroad, 1992), p. 8.

10 *Transcendence and Spatiality of the Triune Creator*

not necessarily less inadequate or improper than old ones."[29] Thus, like other feminist theologians, she holds that, even as "God the father" is also a metaphor, we can freely speak of God as mother as well. McFague says, "Metaphor always has the character of 'is' and 'is not': an assertion is made but as a likely account rather than a definition."[30] In relation to metaphor, she asserts, a model is a metaphor with "staying power" or "sufficient stability and scope so as to present a pattern for relatively comprehensive and coherent explanation."[31]

Although feminist theologians such as Johnson and McFague argue that Christian theology has a metaphorical character and therefore is amenable to new metaphors, their contention may be criticized from other perspectives. It may be that changing some metaphors can lead to descriptions of other gods rather than the triune God of the Scriptures. For Colin Gunton and John Cooper, metaphors for God are not only figurative but can also truly describe. Gunton and Cooper oppose the feminist position that "Language for God is figurative and cannot truly describe."[32]

In this book, I will argue that God is, figuratively or metaphorically, a space for us. As references for this contention, I will refer to some biblical verses which speak of God as our dwelling place (e.g., Ps. 90:1; Deut. 33:27; Acts 17:28). Thus, I will adopt the option that God is space in a metaphorical sense. Although I dispute the view that God has space as a sense organ, I will argue that God has his own space different from that of creatures. God can also use the limited space of this world as a vehicle, or medium, of his creative presence. Even though God is the transcendent Creator of space and time of the universe, he can be present in limited space by means of a relational concept of space: God is present in space.

McFague contends that to speak of the world as God's body is a kind of metaphor. Quoting Exodus 33:23 "And you shall see my back; But my face shall not be seen," McFague suggests, "This body is but the backside of God, not the face; it is the visible, mediated form of God,

---

29　Sallie McFague, *Models of God: Theology for an Ecological, Nuclear Age* (Philadephia: Fortress, 1987), p. 35.

30　McFague, *Models of God*, p. 33.

31　McFague, *Models of God*, p. 34.

32　John W. Cooper, "Synodical Report of Committee to Study Inclusive Language for God" (unpublished report), p. 34f.

Introduction 11

one that we are invited to contemplate for intimations of divine transcendence."[33] In this sense, McFague's view is less radical than that of Grace Jantzen who contends for the world as the body of God.[34] However, McFague's position is not acceptable to Christian theology, either. McFague holds that "the idea of God's embodiment [...] should not be seen as nonsense; it is less nonsense than the idea of a disembodied personal God."[35] In relation to the issue of God's embodiment, McFague explains pantheism, traditional theism, and panentheism, respectively, as follows: "Pantheism says that God is embodied, necessarily and totally; traditional theism claims that God is disembodied, necessarily and totally; panentheism suggests that God is embodied but not necessarily or totally. Rather, God is sacramentally embodied: God is mediated, expressed, in and through embodiment, but not necessarily or totally."[36] For McFague then, the world (universe) is God's body and God is its spirit. McFague's speaking of the world as God's body is more than a metaphor; it is a model.

In opposition to McFague, Thomas Tracy argues for the disembodied God's action in the world. Tracy's main opponent is process theology. Yet, even though he opposes the idea of the embodiment of God, Tracy does not seem to deny the second category of Dyrness, embodiment. For, with this category, Dyrness does not support the model of the world as God's body. Nor does Tracy oppose to Dyrness' category of embodiment.

This book opposes the view of the so-called "God of the gaps." I will maintain here that there is the space, room or "gaps" for God's action in the world. Since Dietrich Bonhoeffer has pointed out the flaws of the "God of the gaps" in one of his later prison letters,[37] the "God of the

33 McFague, *The Body of God*, p. 156.
34 Grace Jantzen, *God's World, God's Body* (Philadelphia: Westminster, 1984).
35 McFague, *Models of God*, p. 71.
36 McFague, *The Body of God*, p. 149f. Elsewhere she presents the following definition of panentheism: "it is a view of the God-world relationship in which all things have their origins in God and nothing exists outside God, though this does not mean that God is reduced to these things" (ibid., 72).
37 Dietrich Bonhoeffer, *Letters and Papers from Prison*, ed. Eberhard Bethge, enlarged ed. (New York: Macmillan, 1979), p. 311, writes as follows: "It has again brought home to me quite clearly how wrong it is to use God as a stop-gap for the incompleteness of our knowledge. If in fact the frontiers of knowledge are being pushed further and further back (and that is bound to be the case), then God is be-

12 *Transcendence and Spatiality of the Triune Creator*

gaps" has not fared well in modern theology. John Polkinghorne has defined "God of the gaps" as follows: "The invocation of God as an explanation of last resort to deal with questions of current (often scientific) ignorance. ('Only God can bring life out of inanimate matter,' etc.)."[38] Most contemporary theologians refute this idea of God of the gaps. Since such a notion of deity is subject to continual decay with the advance of knowledge, it is theologically inadequate. Not only Polkinghorne but also Arthur Peacocke and Nancey Murphy are the representative theologians who oppose the God of the gaps.[39] According to them, this "God of the gaps" cannot be the transcendent creator of the world.

In agreement with Polkinghorne, Peacocke, and Murphy, this study opposes the view of the "God of the gaps." Unlike these three thinkers, however, I will contend for the "gaps" for God's action in the world. In his article, "Particular Providence and the God of the Gaps," Thomas F. Tracy makes a useful observation on this point: "God would be the creator not only of natural law but also of the indeterministic gaps through which the world remains open to possibilities not exhaustively specified by its past."[40] Following Tracy, though not supporting the "God of gaps," I will use the term "gaps" in a neutral sense. Some theologians, such as Friedrich Schleiermacher, would suggest the possibility that, while God cannot act in history, God does enact the history of the world. Yet, this

---

ing pushed back with then, and is therefore continually in retreat. We are to find God in what we know, not in what we don't know."

38 John Polkinghorne, *The Faith of a Physicist: Reflections of a Bottom-Up Thinker* (Minneapolis: Fortress Press, 1996), p. 197.

39 Concerning Arthur Pecocke's view, see Peacocke, *Intimations of Reality: Critical Realism in Science and Religion* (Notre Dame: Notre Dame University Press, 1984), p. 57, Peacocke, *Theology for a Scientific Age: Being and Becoming – Natural, Divine, and Human* (Minneapolis: Fortress Press, 1993, Enlarged Ed.), p. 144f, and Peacocke, "God's Interaction with the World: The Implications of Deterministic 'Chaos' and of Interconnected and Interdependent Complexity," in R. J. Russell, N. Murphy, and A. R. Peacocke, eds, *Chaos and Complexity: Scientific Perspectives on Divine Action* (Vatican City State: Vatican Observatory Publications, 1997, 2nd Ed.), p. 277. And concerning Nancey Murphy's position, see Murphy, "Divine Action in the Natural Order," R. J. Russell, N. Murphy, and A. R. Peacocke, eds, *Chaos and Complexity: Scientific Perspectives on Divine Action*, p. 343.

40 Thomas F. Tracy, "Particular Providence and the God of the Gaps," R. J. Russell, N. Murphy, and A. R. Peacocke, eds, *Chaos and Complexity: Scientific Perspectives on Divine Action*, p. 294.

Introduction   13

position is in fact not distinguishable from deism. It does not admit God's place in the universe. In other words, it does not acknowledge space, room or gaps for God's action in this world. However, the world that is known to us through contemporary natural science is not closed but open-ended. It is not possible to explain various phenomena of this world completely without indeterministic gaps. Within the universe, there are some gaps. These gaps are not just epistemological, as in the view of the God of the gaps, but are ontological or intrinsic to this universe.

The purpose of the present book is to make two contributions to Christian theology. First, it will present a discussion and defense of God's spatiality. This study will be argued in a way similar to the approach of Ted Peters' work on the temporality of God. In the last chapter of his book, *GOD as Trinity*, Ted Peters presents his view on God's temporality. He converses with the new discoveries of the contemporary natural sciences: relativity theory, quantum physics, and the work of Stephen Hawking. He comes to his own conclusion much of which is adopted from the thought of Pannenberg. For me, whether or not Peters' conclusion is right, his attempt itself seems to be valuable. In a way similar to Peters', my study will attempt to present a view on God's spatiality.

As a second contribution, this book will attempt to criticize and correct for the influence of gnosticism in traditional theology. "Gnosticism" has been well defined David F. Ford: "Name (derived from the Greek word *gnosis*, knowledge) given to a varied and diffuse religious movement which saw creation as the work of an inferior god, sharply divided the physical from the spiritual, and offered to initiate exclusive knowledge enabling them to escape from the world and physicality to union with the supreme divine being."[41] Since the early church, the gospel of Christianity has been influenced and threatened by a wide variety of gnostic-like heresies. A corollary result of this influence is that Christians tend to despise the bodily existence of human beings in this world as a result of an unbalanced other-worldliness. This other-worldly spirituality is not to be abandoned, but rather to be complemented, or balanced, by an adequate emphasis on the bodily life in this world. On this issue, the above mentioned book of Dyrness has made a significant con-

---

41   Ford, ed., *The Modern Theologians*, p. 741f.

14   *Transcendence and Spatiality of the Triune Creator*

tribution. I hope my book will further defend and expand on the theological ground underlying his book.

3. Thesis and Plan

The thesis of this study may be stated as follows. Divine transcendence does not necessarily exclude divine spatiality. In fact, there are distinct biblical and theological evidences for God's spatiality. Even though God may be interpreted as radically different from creatures, Christians must believe that in some sense God is spatial. For God is relational, both in himself and with the created world. God interacts with creation and with us within the world of space and time. Although He does not have His own body, since He embodied Himself in Jesus Christ, God should be considered in some sense spatial. God is figuratively space and has His own space other than that of creatures. Even though He is transcendent over the world of space and time, God can be present in the space as well as the time of this world. To describe such a co-presence of the infinite and the finite, we do need a relational, or relative, concept of space, not the Newtonian absolute space.

The body of this book is divided into five chapters. *Chapter II* will present an historical survey of the development of the concept of space with respect to Christian theology. This chapter will show that the old receptacle idea of space proves troublesome to account for much of reality. Besides the absolute space of Newton and Kant, I will further argue, we need a relational, or relative, space that is also consistent with relativity theory. Among contemporary theologians, Torrance and Moltmann, have developed views of space that comport with contemporary epistemologies. The former has identified the concept of space of the Nicene Fathers and the Reformed theologians with that of contemporary physicists, including Einstein. The latter has demonstrated the limits of the geometric concept of space. In developing his doctrine of creation, therefore, Moltmann has asserted an ecological concept of space. Furthermore, in his eschatology he has proposed a perichoretic concept of space. It is neither an absolute nor a relational one. It is, rather, a kind of divine space, which offers a deepened model of relational space.

Introduction 15

Chapter III begins by defending the doctrine of the divine transcendence and by suggesting the pitfalls of panentheism. Panentheism, I will contend, blurs the distinction between God and the world and logically leads to a pantheistic view. Panentheism fails to uphold the doctrine of divine transcendence. This chapter further describes the appearance of deism as a position which emerged with the development of modern science. It upheld God's absolute transcendence, but it provided for no discussion of God's interaction with the world. Thus deism was eventually rejected as a traditional option. Beyond the apophatic position of Gunton, I will explore various possibilities for talking about God's spatiality which is not contradictory to the divine transcendence but comports with the biblical and theological data.

Chapter IV deals with Torrance, Pannenberg, and Moltmann in relation to God and his spatiality. Torrance's idea of the redemption of space is parallel to that of time. While he admits the temporality of God, Torrance does not overtly speak of God's spatiality. Yet, it seems to me, that he should have explored for it. Similar to Torrance, Pannenberg explains the Spirit in terms of a force field. This idea has the advantage of explaining God's presence in this world. However, it cannot escape the charge of being panentheistic because it also fails to retain a personal character of God. Unlike Torrance and Pannenberg, Moltmann openly supports the position of panentheism. He explains God's creation of the world as a divine *zimzum*, i.e., self-contraction. While Torrance has probably avoided contending for the spatiality of God, because he is concerned with the divine transcendence, Pannenberg and Moltmann have opted for panentheism. These outcomes would suggest that no way has been found to explain God's spatiality apart from "panentheism." The present dissertation will argue to the contrary, that such a solution is possible.

Chapter V will examine some options for speaking of God's spatiality. It will present several developments for accounting for divine spatiality. In his recent book Moltmann boldly avers that God is space for us creatures. I will argue this view is mistaken because it identifies the divine space with the triune God himself. Christians should not take Psalm 90:1 literally: "You have been our dwelling place throughout all generations." Rather, Christians should speak of God as the indwelling place for us, but this "place" is a metaphorical space. The representative of the option that God has a kind of space is Newton's absolute space as

16 *Transcendence and Spatiality of the Triune Creator*

sensorium Dei. Even though it is controversial as to whether or not Newton was referring to the divine sensory with the term *sensorium Dei*, we can discard the option that God has space as something like his sense organ. Another option is to speak of God as having his own space as well as his own time. But what does it mean to say that God has space? The option that God has his own space is viable, only when this question is clarified. On the creaturely level, to have space means to have a bodily existence. The abhorrence of body in Christian tradition has a long history. It is due to the influence of Greek philosophy and the misunderstanding of God's spirituality. The central orthodox message of Christianity about the future of human beings is the bodily resurrection. Especially in Reformed tradition, the ascended Christ still has humanity, which includes in some sense his body. Even though we cannot give assent to the bold position of Sallie McFague, that this world is God's body, we should take our bodily life in the world seriously. In this sense, *The Earth is God's* gives us some significant insight. Dyrness tries to explain the importance of our bodily lives through three normative categories, which are grounded in the triune God's presence. He asserts that, "Spatial and tactual dimensions must be added to our intellectual map and our spiritual journey."[42] Tracy argues that it is possible to explain God's action without his embodiment. Since God is a perfect agent, Tracy opposes the idea of divine embodiment. Thus Tracy rejects the notion of divine embodiment as set forth by process theologians and Sallie McFague, but Tracy's position would not deny the second category of Dyrness. Tracy also admits some kind of divine embodiment. By means of a relational concept of space, I contend that God is present in the limited space of this world. In this context, the term perichoresis gets the focus of my discussion.

Finally, *Chapter VI* will deal in detail with three categories set forth by Dyrness: relationship, agency and embodiment. All of these categories are grounded in the triune presence of God. These grow out His own trinitarian presence and reality. Each section of this chapter develops these categories respectively. We will make use of contemporary discussions: Gunton and Peters assert that God's relationality is a key theme in

---

42   Dyrness, *The Earth is God's*, p. 160.

Introduction

today's Trinity talk.[43] Catherine LaCugna is also concerned with the relational character of the divine life.[44] In LaCugna's view, God is relational both in himself and with the created world. Discussion on God's relationality is possible only when we presuppose a kind of divine spatiality. Peacocke contends that chance is the radar of God. He sees chance in biology as the means of God's action in the world. Murphy proposes a way of divine action to avoid both extremes of interventionism and immanentism.[45] God interacts in the world of space and time, especially in the quantum world and through human intelligence. As an alternative to the quantum indeterminacy, Polkinghorne proposes chaos theory as the space for God's action in this universe. Discussion on divine action is possible when we presuppose a kind of divine spatiality. God has created the universe. Creation is the theatre for God's glory. God does not have his own body, but he embodied himself in various ways and finally in Jesus Christ. The church is the body of Christ. God is present in the Eucharist. New creation, which makes creation a suitable vehicle for God's glory, is an embodied event. Discussion on God's embodiment is possible when we presuppose a kind of divine embodiment. Through these three categories of relationship, agency, and embodiment, I will try to explain the presence of the triune Creator God in our world and present three positive meanings of God's spatiality. Without divine spatiality as a presupposition, it is not possible to explain the presence of God clearly and correctly.

---

43   See Colin E. Gunton, "Relation and Relativity: The Trinity and the Created World" Christopher Schwöbel ed., *Trinitarian Theology Today: Essays on Divine Being and Act* (Edinburgh: T&T Clark, 1995) and Ted Peters, *God as Trinity: Relationality and Temporality in Divine Life* (Louisville, KN: Westminster/John Knox Press, 1993).

44   Catherine M. LaCugna, *God for Us: The Trinity and Christian Life* (New York: Harper Collins Publishers, 1991).

45   Nancey Murphy, *Beyond Liberalism and Fundamentalism: How Modern and Postmodern Philosophy Set the Theological Agenda* (Valley Forge, PA: Trinity Press International, 1996).

Chapter II

Concepts of Space

1. Introduction

In his *Confessions* (XI.14.17), Augustine opens his famous investigation of the nature of time with the following saying: "When nobody is asking me, I know what it is, but when I try to explain it to somebody who asks me, then I don't know."[1] We can also ask this question: what is space? When we ask this question, we might feel the same perplexity as Augustine felt for time. Augustine argues that, since time is also created, time begins with God's creation. While creation is temporal, God is eternal. God is an atemporal being.

In our human experience, space and time are not symmetrical: "we can experience different times in the same space, but not different spaces at the same time."[2] Space and time are complementary and inseparable in the space-time continuum of physics. Furthermore, in comparison to time, space arouses little attention in the areas of theology as well as philosophy. Thus, Moltmann laments, "I found many theological and philosophical studies of time, but almost none of space. The only standard work is Max Jammer's book, with a foreword by Albert Einstein."[3] Dyrness comments on the same situation as follows:

---

1   Cited by Wolfhart Pannenberg, "Eternity, Time, and the Trinitarian God," Colin E. Gunton ed., *Trinity, Time, and Church: A Response to the Theology of Robert W. Jenson* (Grand Rapids: Eerdmans, 2000), p. 62. Moltmann also refers to this saying of Augustine. See Jürgen Moltmann, *God in Creation* (Minneapolis: Fortress Press, 1993), p. 104.

2   Moltmann, *Experiences in Theology: Ways and Forms of Christian Theology* (Minneapolis: Fortress, 2000), p. 315.

3   Moltmann, *Experiences in Theology*, p. 315. To my regret, Moltmann does not refer to T. F. Torrance, *Space, Time and Incarnation* (Edinburgh: T&T Clark, 1969).

20   *Transcendence and Spatiality of the Triune Creator*

> A Theology of history and narrative then needs a correction and a deepening. How can this be characterized? At the risk of oversimplification we might put matters in this way. History and story are categories that readily incorporate time but not *space*. They imply movement, as in a melody, but they cannot fully comprehend the soil – the complexity in which this movement is rooted.[4]

History implies the dimensions both of time and space. In theology of history, however, time alone has been the focus. In theology of story, the so-called *Sitz im Leben* of the community which speaks a story, has little dealt with in terms of space in comparison with the attention given to time. It should include the dimensions of both time and space.

This chapter presents a historical survey of the various concepts of space and then introduces the views of two modern theologians. There are two rival concepts of space: the receptacle notion of space and the relational concept of space. Patristic theology rejected the former but accepted the latter (Section 2). By the notion of absolute space, Newton sought to speak of space as an infinite receptacle in terms of the infinity of God. Kant understood space as *a priori* forms of intuition along with time. Leibniz rejected Newton's conception of absolute space (Section 3). Leibniz's relational or relative concept of space was finally justified, in a roundabout way, in the relativity theory of Einstein, who observed that space is not an independent reality from the universe (Section 4). Torrance has presented a Reformed view on space. He identifies the concept of space in Patristic theology and Reformed theology with space as it is treated in relativity theory, that is, as a relational, relative, or dynamic notion (Section 5). Moltmann has differentiated the concepts of space into three groupings: geometrical, ecological, and perichoretic concepts of space. He develops and pushes too far a perichoretic concept of space in his eschatology. However, Moltmann has not seemed to distinguish the relational concept of space from the receptacle one (Section 6).

---

4   W. Dyrness, *The Earth is God's: A Theology of American Culture* (Maryknoll, NY: Orbis Books, 1997), p. 13.

Concepts of Space   21

2. Two Concepts of Space

In the Foreword to Max Jammer's book *Concepts of Space*, Einstein contrasts two concepts of space: (a) space as a positional quality of the world of material objects; (b) space as a container of all material objects. In case (a), space does not have its own independent existence apart from material objects. Without a material object, space itself is inconceivable. This concept of space seems to be preceded by the psychologically simpler concept of "place." Einstein states, "Place is first of all a (small) portion of the earth's surface identified by a name. The thing whose 'place' is being specified is a 'material object' or body."[5] In this case, therefore, to speak of empty space has no meaning.

In case (b), however, space has its own existence apart from material objects, which can only be conceived as existing in space. In this case, space is in a certain sense superior to the material world and has its own independent existence. While a material object not situated in space is simply inconceivable, it is quite conceivable that an empty space may exist. Einstein calls this type of space "box space."[6] More popularly, this view has been referred to as "a receptacle or a container notion of space."[7] This concept of space is consistent with that which dominated ancient Greek thought.

Thomas F. Torrance further classifies receptacle notions of space as finite and infinite. The finite receptacle concept of space entered into Western thought with Aristotelian physics and philosophy, under the category of magnitude, apart from the conception of time. Torrance criticizes this notion of space: "it tended to encourage a *deus sive natura* view of the relation between God and the world but at the same time led to their separation in a sharp distinction between nature and supernature and consequently to a corresponding distinction between natural theology and revealed theology."[8] Torrance comments that patristic theology rejected this notion of space.

---

5   Max Jammer, *Concepts of Space: The History of Theories of Space in Physics* (New York: Dover Publishings, Inc., 1993, Enlarged Ed.), p. xv.

6   Jammer, *Concepts of Space*, p. xv.

7   Torrance, *Space, Time and Incarnation*, p. 4.

8   Torrance, *Space, Time and Incarnation*, p. 56.

22 *Transcendence and Spatiality of the Triune Creator*

The infinite receptacle notion of space entered Renaissance thought through the Florentine Academy. It is evident in the thinking of people as different as Patritius, Gassendi and Galileo. Newton later gave space together with time an absolute status independent of material bodies but causally conditioning their character as an inertial system, thus making nature determinate and our knowledge of nature possible. "Since space and time were regarded by Newton as attributes of God, constituting the infinite Container of all creaturely existence," Torrance states, "this view also tended towards a *deus sive natura* conception of the relation between God and the world by giving them space and time in common."[9] Thus he comments,

> Be that as it may, throughout history the receptacle notion of space has proved very troublesome for theology, but perhaps never more than in modern times when space came to be thought of as a container independent of what takes place in it and regarded as an inertial system exercising an absolute role in the whole causal structure of classical physics.[10]

According to Torrance, therefore, both the finite receptacle notion of space and the infinite notion are to be rejected in Christian theology.

Instead of the receptacle notion of space, Torrance maintains that Patristic theology developed a notion of space as the seat of relations or the place of meeting and activity in the interaction between God and the world. This notion of space was brought to its sharpest focus in Jesus Christ. For Jesus Christ is "the place where God has made room for Himself in the midst of our human existence" and "the place where man on earth and in history may meet and have communion with the heavenly Father."[11] As they had to develop their thought in accordance with the nature of the creator who transcends all space and time and in accordance with the nature of the creature subject to space and time, the church fathers essentially needed a "differential and open concept of space" sharply opposed to the Aristotelian idea of space as the immobile limit of the containing body.[12] This relational idea of space was later given its supreme expression in the space-time of Einstein's relativity

9   Torrance, *Space, Time and Incarnation*, p. 57.
10   Torrance, *Space, Time and Incarnation*, p. 22.
11   Torrance, *Space, Time and Incarnation*, p. 24.
12   Torrance, *Space, Time and Incarnation*, p. 24f.

Concepts of Space   23

theory. An elaboration of this concept of space will be given below, after first dealing with the controversial aspects of Newton's absolute space.

3.   Controversy over Absolute Space

Einstein explains the concept of absolute space in relation to the concept of "box space" as follows: "By a natural extension of 'box space' one can arrive at the concept of an independent (absolute) space, unlimited in extent, in which all material objects are contained."[13] In other words, we may see the notion of absolute space as the infinite receptacle notion of space. In this section, the theory of absolute space, as it finally crystallized in Newtonian mechanics, will be examined together with the criticism of it by the first modern relativists, Leibniz and Huygens.[14]

In contrast with Leibniz and Huygens, it was clear to Newton that the space concept of type (a) which is stated above, that is, space as positional quality of the world of material objects, was not sufficient to serve as the foundation for the inertia principle and the laws of motion.[15] In Newtonian thinking, according to Jammer, absolute space is a logical and ontological necessity. It is a necessary prerequisite for the validity of the first law of motion: "Every body continues in its state of rest, or of uniform motion in a right line, unless it is compelled to change that state by forces impressed upon it."[16] The state of rest also presupposes such an absolute space. In short, absolute space is indispensable to Newtonian mechanics.

In Newton's conception of absolute space there is a synthesis of two heterogeneous elements: "One of these elements is rooted in the emancipation of space from the scholastic substance-accident scheme, a scheme which was finally abandoned by the Italian natural philosophers of the

---

13   Jammer, *Concepts of Space*, p. xv.
14   Cf. Jammer, *Concepts of Space*, p. 2.
15   Jammer, *Concepts of Space*, p. xvi.
16   F. Cajori, ed., *Sir Isaac Newton's Mathematical Principles of Natural Philosophy and His System of the World: A Revision of Mott's Translation* (Berkeley: University of California Press, 1934), p. 13. This passage is cited in Jammer, *Concepts of Space*, p. 101.

24 *Transcendence and Spatiality of the Triune Creator*

Renaissance. The other element draws on certain ideas that identify space with an attribute of God."[17] In scholastic Aristotelianism, space is identified with place and defined as the adjacent boundary of the containing body. This definition of space is in line with Aristotle's fundamental assumption of the impossibility of a vacuum: "Since the containing body is always in immediate limiting contact with what is contained, there can be no void or empty space."[18] In this Aristotelianism, space is not an independent substance but a kind of accident. Hence, space does not have its own independent existence. Aristotle's view of space, or place, as the surface of the *adjacent* body, went unchallenged for a long time.

Tolesio was the first to suggest another interpretation. He adopted certain materialistic and Stoic conceptions of Antiquity, which led him to ascribe to spiritual functions a certain degree of corporeality. Jammer observes, "Space ceases with Tolesio to be a mere quality and assumes an independent existence. Space is the great receptor of all being whatever."[19] In Tolesio's view, space is not a quality of reality. Rather, qualities themselves are dependent on space. Thus, for Tolesio, space does not fit into the substance-accident scheme. Gassendi pushed Tolesio's thought a little further, stating that "space is neither a mode nor an attribute; both of these exist in subordination to the object to which they belong, whereas space is independent of any substance."[20] For Gassendi, space is infinite, unchangeable and immovable. According to Jammer, "it was Newton who incorporated Gassendi's theory of space into his great synthesis and placed it as the concept of absolute space in the front line of physics."[21]

We should look into one more background factor of Newton's conception of absolute space. Newton was influenced by Henry More, an ardent scholar of cabalistic lore. In order to substantiate his vigorous religious beliefs, More tried to augment the science of Descartes with cabalistic and Platonic concepts. "As far as his theory of space is concerned, More himself refers to the cabalistic doctrine as explained by Cornelius Agrippa in his *De occulta philosophia*, where space is speci-

17   Jammer, *Concepts of Space*, p. 2.
18   Torrance, *Space, Time and Incarnation*, p. 8.
19   Jammer, *Concepts of Space*, p. 85.
20   Jammer, *Concepts of Space*, p. 93.
21   Jammer, *Concepts of Space*, p. 94.

Concepts of Space   25

fied as one of the attributes of God."[22] Jammer summarizes More's treatment of the space problem according to three propositions: (1) extension is not the distinguishing attribute of matter; (2) space is real, having real attributes; (3) space is of divine character.[23] Significantly, for the interest of the present study, Jammer connects More's conception of space to Luria's cabalistic notion of the *"zimzum,"* the divine self-concentration, creating space by self-restriction: "A somehow pantheistic interpretation of the cabala must necessarily lead to More's conception of space."[24]

Newton separated space from what happened in it and gave it an absolute status independent of material existence. This concept of absolute space is necessary for making nature determinate and our knowledge of nature possible. Newton spoke of space and time as an infinite receptacle in terms of the infinity and eternity of God. Infinite space and time are in fact attributes of Deity.[25] For Torrance, the fact that Newton associated space and time as an infinite receptacle with Deity had the effect of reinforcing the dualism between space and matter: "If God Himself is the infinite Container of all things He can no more become incarnate than a box can become one of the several objects that it contains."[26] Thus, Torrance observes, "Newton found himself in sharp conflict with Nicene theology and its famous *homoousion*, and even set himself to defend Arius against Athanasius."[27]

Pannenberg seems to support Newton's conception of absolute space. He indicates the possibility of regarding overt attempts to counter the dominance of Newtonian mechanism, such as Faraday's field theory, as an underlying renewal of the deeper intentions of Newton himself. Nevertheless, like Torrance, Pannenberg also sees a fatal weakness of Newton's theology. Pannenberg emphasizes the importance of the doctrine of the Trinity in a Christian theology of creation: "Newton's theology was not pantheistic, as was thought, but it had also no relationship to the doc-

---

22   Jammer, *Concepts of Space*, p. 41.
23   Jammer, *Concepts of Space*, p. 43.
24   Jammer, *Concepts of Space*, p. 48f. Moltmann accepts this idea of divine *zimzum* in his doctrine of creation. See Moltmann, *God in Creation* (Minneapolis: Fortress Press, 1993), p. 87f.
25   Torrance, *Space, Time and Incarnation*, p. 38.
26   Torrance, *Space, Time and Incarnation*, p. 39.
27   Torrance, *Space, Time and Incarnation*, p. 39f.

26   *Transcendence and Spatiality of the Triune Creator*

trine of the Trinity. Today's Christian theology of creation will use, in distinction from Newton, the possibilities of the doctrine of the Trinity in order to describe the relationship of God's transcendence and immanence in creation and in the history of salvation."[28]

Concerning Newton's idea of absolute space as the *sensorium Dei*, his contemporary Leibniz leveled strong criticism against this conception of space (and time), claiming that it tended to slip into pantheism. Concerning this issue, Pannenberg has noted,

> Newton's well-known conception of space as sensory of God *(sensorium Dei)* did not intend to ascribe to God an organ of perception, the like of which God does not need, according to Newton, because of divine omnipresence. Rather, Newton took space as a medium of God's creative presence at the finite place of his creatures in creating them. The idea of Newton was easily mistaken as indicating some monstrously pantheistic conception of God similar to that found in Leibniz's polemics against Newton.[29]

On the one hand, Einstein treated the attack of Leibniz as "supported by inadequate arguments." Yet, on the other hand Einstein asserted that Leibniz's relational concept of space was "intuitively well founded."[30] Leibniz rejected Newton's conception of an absolute space, on the ground that space is nothing but a network of relations among coexisting things. Thus Leibniz's relational concept of space was actually justified in a roundabout way throughout the subsequent development of the problems which no one then could possibly foresee.[31]

4. Concept of Space in the Theory of Relativity

Max Jammer indicates Kant's temporary attempt to reconcile Newton with Leibniz:

---

28   Pannenberg, *Toward a Theology of Nature*, p. 65.

29   Pannenberg, *Toward a Theology of Nature*, p. 42.

30   Jammer, *Concepts of Space*, p. xvi. I will examine the controversy between Leibniz and Clarke over the concept of absolute space as the *sensorium Dei* in the Section 3 of Chapter V.

31   Jammer, *Concepts of Space*, p. xvi.

Concepts of Space

27

Agreeing with Leibniz's relational point of view, Kant sees in spatial relations not reflections of simply qualitative data given within the order of existing matter but rather mutual effects and interactions among bodies; since causal interdependence is not given with matter as such, but has been added and imparted by divine creation, space is an independent existent of absolute reality in the Newtonian sense.[32]

But Kant finally abandons this point of view and declares himself in favor of the Newtonian concepts of absolute space and absolute time.

In his famous book, *Critique of Pure Reason*, Kant understands space and time as "two pure forms of sensible intuition." Space and time serve as principles of *a priori* knowledge. For Kant, space is not "an empirical concept which has been derived from outer experiences," but rather "a necessary *a priori* representation, which underlies all outer intuitions."[33] Space is neither a property of things in themselves nor a relation of things to one another. Kant claims, "Space is nothing but the form of all appearances of outer sense. It is the subjective condition of sensibility, under which alone outer intuition is possible for us."[34] In distinction from Newton's "absolutist" view and Leibniz's "relational" view, thus, it is possible to express Kant's view on space and time as follows: space and time are not "only determination or relations of things" but "real existences." Kant supports the Newtonian view. Sebastian Gardner contends that "Newton's view is of space as an absolutely real, self-subsistent 'container' which would exist even if no physical objects were contained within it."[35]

Jammer has described the situation after Newton. "The notion of absolute space triumphed on all fronts."[36] Prior to Newton, for Aristotle, space was an accident of substance; following Newton, substance was seen as an accident of space. By the notion of absolute space, "the concept of space, after its emancipation during the Renaissance, seized totalitarian power in a triumphant victory over the other concepts in theoretical physics."[37]

---

32 Jammer, *Concepts of Space*, p. 131.
33 Immanuel Kant, *Critique of Pure Reason*, trans. Norman Kemp Smith (New York: The Humanities Press, 1950), p. 68.
34 Kant, *Critique of Pure Reason*, p. 71.
35 S. Gardner, *Kant and the Critique of Pure Reason* (London and New York, Routledge, 1999), p. 70.
36 Jammer, *Concepts of Space*, p. 129.
37 Jammer, *Concepts of Space*, p. 163.

28   *Transcendence and Spatiality of the Triune Creator*

Ian Barbour describes the effect of Einstein's theory of relativity on Newton's view as follows:

> For Newton and throughout classical physics, space and time are separable and absolute. Space is like an empty container in which every object has a definite location. Time passes uniformly and universally, the same for all observers. The cosmos consists of the total of all such objects in space at the mass of an object are unchanging, intrinsic, objective properties, independent of the observer. All of this is close to our everyday experience and common-sense assumptions, but it is challenged by relativity.[38]

By the special theory of relativity, Einstein postulates "the constancy of the velocity of light for all observers." The result of this hypothesis, Barbour states, is that, "The two events are simultaneous in one frame of reference but not in the other. The effect would be very small with a train but would be large with a space rocket or a high-energy particle approaching the velocity of light."[39] In special relativity, Barbour explains, "Space and time, then, are not independent but are united in a *spacetime continuum*. The spatial separation of two events varies according to the observer, and the temporal separation also varies, but the two variations are correlated in a definite way."[40] By means of the general theory of relativity, Einstein extends his ideas to include gravity. Concerning this part of Einstein's thought, Barbour explains:

> He [Einstein] reasoned that an observer in a windowless elevator or space ship cannot tell the effects of a gravitational field from the effects of accelerated motion. From this he concluded that the geometry of space is itself affected by matter. Gravity bends space, giving it a four-dimensional curvature (here the fourth dimension is spatial rather than temporal, and is reflected in the altered geometry of

---

38   Ian Barbour, *Religion and Science: Historical and Contemporary Issues* (New York: HarperCollins Publishers, 1997), p. 177.

39   Barbour, *Religion and Science*, p. 178. He explains like this: "Imagine that an observer at the middle of a moving railway train sends light signals, that reach the equidistant front and rear of the train at the same instant. For an observer on the ground, the signals travel different distances to the two ends (since the train moves while the signals are traveling); therefore if the signals travel at constant velocity in his framework they must arrive at different times" (ibid.).

40   Barbour, *Religion and Science*, p. 178.

Concepts of Space

three-dimensional space). As John Wheeler puts it, "Space tells matter how to move, and matter tells space how to curve."[41]

In the general theory of relativity, space is curved by gravity and time is also shrunk by it. By the theory of relativity, the concepts of absolute space and time are challenged and discarded.

After referring to the relation of special relativity to time,[42] Arthur Peacocke also indicates the impact of general relativity to time. Peacocke states, "Also of general significance is the renewed support that the theory of general relativity now gives to the idea, already noted, in accord with St Augustine's famous assertion, of physical ('clock') time being an aspect of the created order, for in that theory time is closely interlocked conceptually with space, matter and energy."[43] Significantly, Peacocke here connects the time theory of Augustine to that of Einstein's theory of relativity.

Examining Augustine's doctrine of creation, Gunton extends Peacocke's view on time to suggest an application to space. According to Greek thought, the demiurge at some time shaped matter and form into *this* particular universe. However, this view is not acceptable to Christian theology. Gunton states that "the Christian view entailed an absolute beginning, and to avoid the embarrassing suggestion that God is in some

---

41 Barbour, *Religion and Science*, p. 179.
42 Arthur Peacocke, *Theology for a Scientific Age: Being and Becoming – Natural, Divine, and Human* (Minneapolis: Fortress Press, 1993, Enlarged Edition), p. 130. Peacocke refers to the relation of special relativity to time: "Special relativity raises a particular difficulty in all talk of God's relation to 'time'. For that theory replaces the one, universal flowing time assumed by Newton and by common sense, by many different 'times' specific to different observers, each with their own positions and velocities, that is, their own frames of reference" (ibid.).
43 Peacocke, *Theology for a Scientific Age*, p. 130. In an endnote, he cites Augustine's saying, "'It is therefore true to say that when you [God] had not made anything there was no time, because time itself was of your making', *Confessions* xi.14, trans. R. S. Pine-Coffin (Penguin Classics edn, Harmondsworth, London, 1961), p. 263" (ibid., p. 364, note 81). Unfortunately, however, Peacocke does not comment on the issue of space in this context. Here is his conclusion on God and time: "*God is not 'timeless'; God is temporal in the sense that the Divine life is successive in its relation to us – God is temporally related to us; God creates and is present to each instant of the (physical and, derivatively, psychological) time of the created world; God transcends past and present created time: God is eternal,* in the sense that there is no time at which he did not exist nor will there be a future time at which he does not exist" (ibid., 132).

30   *Transcendence and Spatiality of the Triune Creator*

way limited by time, Augustine, in anticipation of views contained in Einstein's relativity theory, argued that space and time were created *with* the world."[44] For Augustine and Einstein, then, "space and time are not absolute realities that in some way constrain God, but are the result of there being a created universe."[45]

According to Einstein, the relational theory of Leibniz eliminated the concept of absolute space. Einstein admits this process to overcome the concept of absolute space is probably by no means as yet completed. He argues that "the victory over the concept of absolute space or over that of the inertial system became possibly only because the concept of the material object was gradually replaced as the fundamental concept of physics by that of the field."[46] Einstein conceives the nature of space or "the spatial character of reality" as simply the four-dimensionality of the field. So Einstein summarizes as follows: "There is then no 'empty' space, that is, there is no space without a field."[47]

Einstein's concept of space is a relational notion. He rejects "the notion of absolute space and time both as taught by Kant, for whom they were *a priori* forms of intuition outside the range of experience, and as taught by Newton, for whom they formed an inertial system independent of material events contained in them but acting on them and conditioning our knowledge of the universe."[48] This view is founded on non-Euclidean geometry such as Riemann's geometry. As Jammer states, "Riemann's geometry, in this respect, contrasted with the finite geometry of Euclid,

---

44   Colin E. Gunton, "The Doctrine of Creation," in *The Cambridge Companion to Christian Doctrine*, ed. C. E. Gunton (Cambridge: Cambridge University Press, 1997), p. 149.

45   Gunton, "The Doctrine of Creation," p. 149.

46   Jammer, *Concepts of Space*, p. xvii.

47   Jammer, *Concepts of Space*, p. xvii.

48   Torrance, *Space, Time and Incarnation*, p. 58. Elsewhere, Torrance states, "He [Einstein] dethroned time and space from their absolute, unvarying, prescriptive role in the Newtonian system and brought them down to empirical reality, where he found them indissolubly integrated with its on-going processes. At the same time he set aside the idea of instantaneous action at a distance, but also set aside the existence of ether (still maintained by Lorentz) and all idea of the substantiality of the field (in Faraday's sense). There now emerged the concept of the continuous field of space-time which interacts with the constituent matter/energy of the universe, integrating everything within it in accordance with its unitary yet variable objective rational order of non-causal connections" [T. F. Torrance, *Divine and Contingent Order* (Edinburgh: T&T Clark, 1998), p. 77f].

Concepts of Space

31

can be compared with Faraday's field interpretation of electrical phenomena that formerly had been explained by actions at a distance."[49]

Pointing out three dubious claims based on relativity, Barbour writes, "Science is said to have shown that everything is relative and there are no absolutes, and this has been cited in support of moral and religious relativism."[50] Against this contention that *"relativity supports relativism,"* he asserts, "the claim is dubious even in physics. Many absolutes have been given up (space, time, mass, and so forth), but there are new ones. The velocity of light is absolute, and the spacetime interval between two events is the same for all observers."[51] Barbour states that relativity shows us a *dynamic* and *interconnected* universe. According to him, relativity can help us to imagine God as *omnipresent* yet *superspatial*. In order to explain this idea, Barbour introduces Karl Heim's thought on God and selfhood: "Karl Heim speaks of God and selfhood as other 'spaces' and 'in another dimension'. The same set of events can be differently ordered in different spaces. Spaces are concurrent frameworks with incommensurable dimensions; they permeate each other without boundaries."[52] So, for example, since God is omnipresent and knows all events instantaneously, Barbour maintains, the limitation on the speed of transmission of physical signals between distant points would not apply to God. He states, "God is neither at rest nor in motion relative to other systems."[53]

The relational view of space as set forth by Einstein was anticipated by Leibniz in certain respects, but its early roots may be traced even further back to Plato's *Timaeus*. Theophrastus, the pupil and critic of Aristotle, also regarded space as a system of orderly interrelations between the positions of bodies.[54] Torrance argues that the concept of space of the Greek church fathers in the patristic period was consistent with this relational conception of space.

---

49 Jammer, *Concepts of Space*, p. 158.
50 Barbour, *Religion and Science*, p. 180. Here are three claims: 1. "Time is illusory and events are determined." 2. "Reality is mental." 3. "Relativity supports relativism" (ibid., p. 179f). He casts suspicion on all of them.
51 Barbour, *Religion and Science*, p. 180.
52 Barbour, *Religion and Science*, p. 180f.
53 Barbour, *Religion and Science*, p. 181.
54 Torrance, *Space, Time and Incarnation*, p. 58.

32 *Transcendence and Spatiality of the Triune Creator*

5. A Reformed Perspective on Space

In his book, *Space, Time and Incarnation*, Torrance presents a thorough discussion of the development of the concept of space in the history of Christian theology. Similar to Einstein, he identifies two main conceptions of space in Western thought: the *receptacle notion* of space and the *relational notion* of space.[55] The former notion has certainly predominated in ancient and modern times. It crops up in two different forms: the finite receptacle and the infinite receptacle. The church fathers of the patristic period rejected both types of the receptacle idea of space for they tended towards a *deus sive natura* conception of the relation between God and the world.[56]

The other conception of space is the relational one. This view finds its supreme expression in Einstein's theory of relativity. Einstein rejects the notion of absolute space and time both as thought by Kant and as thought by Newton. The Greek church fathers held a view consistent with the relational conception of space and time. Torrance states,

> In the light of God's creation of the world out of nothing, His interaction with nature, and the Incarnation of His Creator Word, they [Greek church fathers] developed a thoroughly relational conception of space and time in which spatial, temporal and conceptual relations were inseparable, for like Plato they held that the basic problem was more an epistemological than a cosmological one.[57]

For Torrance, without a valid conceptualization of space, there can be no right understanding of the Christian doctrines of creation, incarnation and God's action. By means of the relational concept of space, however, Christian theologians have developed these doctrines.

Plato readily spoke of space as a 'receptacle' (ὑποδοχή), but only in a metaphorical sense. He, however, "rejected the notion of a receptacle that had the power to set limits to the bodies it contained."[58] Plato thought of space as helping in some way to bridge the separation between the realms of the intelligible and the sensible. Aristotle miscon-

---

55 Torrance, *Space, Time and Incarnation*, p. 56.
56 Torrance, *Space, Time and Incarnation*, p. 57.
57 Torrance, *Space, Time and Incarnation*, p. 58.
58 Torrance, *Space, Time and Incarnation*, p. 4.

Concepts of Space

33

strued the Platonic separation as a local or spatial separation, and mistook the Platonic 'receptacle' or 'matrix' for the original stuff or substrate, from which bodies are derived. "The model that Aristotle used throughout in developing his notion of place or space (ὁ τόπος τι καὶ ἡ χώρα) was the vessel (ἀγγεῖον) into which and out of which things pass and which not only contains them but exercises a certain force or causal activity (ἔχει τινα δύναμιν) in relation to them."[59] While this Aristotelian definition of space had no place at all in Nicene theology, Torrance avers, Platonic and Stoic notions were more helpful to the Christian Church.[60]

The Stoics offered a very different approach to space: "body extends and makes room for itself through body. Space was thus conceived not in terms of the limits of a receptacle but in terms of body as an agency creating room for itself and extending through itself, thus making the cosmos a sphere of operation and place."[61] The Stoics thought of God, body, and laws of nature, as given together in and with the determinate content of the universe. Thus, for Stoics, "'God' and 'body' were in fact different ways of speaking of the physical principle of rationality which could appear equally as mind and as matter."[62] Commenting on the influence of this understanding of Stoics, Torrance presents a critical insight from Christian theology:

> [B]y binding the idea of God to a finite universe, conceived as a determinate and limited sphere bounded by the infinite void yet having its center in the earth, the Stoics not only failed to reach, or lost, any understanding of the transcendence of God, but made it too easy for the popular and certainly the pagan mind to confound God with nature, theology with cosmology.[63]

Torrance contends that, along with the relational view of space, the Greek church fathers accepted the Stoic view that space must be understood in terms of an active principle that makes room for itself in the finite universe. They refused the receptacle idea of space on the ground of the biblical teaching that God is the transcendent Source of all being whom the heaven of heavens cannot contain. Therefore, as Torrance

---

59 Torrance, *Space, Time and Incarnation*, p. 7.
60 Torrance, *Space, Time and Incarnation*, p. 11.
61 Torrance, *Space, Time and Incarnation*, p. 9.
62 Torrance, *Space, Time and Incarnation*, p. 10.
63 Torrance, *Space, Time and Incarnation*, p. 10.

34 *Transcendence and Spatiality of the Triune Creator*

explains, the Nicene fathers suggested a relationship between the transcendent God and the spatio-temporal universe as follows:

> The doctrine of creation out of nothing, which very quickly came to the forefront of Christian thought, asserted the absolute priority of God over all time and space, for the latter arise only in and with created existence and must be conceived as relations within the created order [...] God Himself, then, cannot be conceived as existing in a temporal or spatial relation to the universe [...] God is not contained by anything but rather that He contains the entire universe, not in the manner of a bodily container, but by His *power* [...] in His own transcendent way God is everywhere and in all things. It follows from this that space and time, and indeed all the structured relations within the universe, have to be understood *dynamically*, through reference to the creative and all-embracing power and activity of God.[64]

God's transcendence as the absolute priority of God over all time and space does not exclude his interaction within the spatio-temporal universe. Yet, this understanding is possible only through the relational or dynamical concept of space.

At the Reformation, according to Torrance, this Patristic idea of space was appropriated by Reformed and Anglican theology. In the further development of Protestant theology, however, this relational conception of space was smothered by the dominance of Newtonian and Lutheran thought which adopted the receptacle notion of space. Since the receptacle notion of space and time has broken down with the advent of the relativity theory, Torrance has proposed "to rethink the essential basis of Christian theology in the relation of the Incarnation to space-time, and to think completely away the damaging effects of a deistic relation between God and the universe."[65]

"Reformed and Anglican theology," Torrance states, "stood much closer to the thought-forms of classical Patristic theology, although it was set out consciously *vis-à-vis* the problems that arose out of debates with the Lutherans regarding the self-emptying of Christ and the ubiquity of His body."[66] Torrance indicates three points of difference between the Lutherans and the Reformed, viz. the so called *"extra Cavinisticum,"*

---

64 Torrance, *Space, Time and Incarnation*, p. 11f.
65 Torrance, *Space, Time and Incarnation*, p. 59.
66 Torrance, *Space, Time and Incarnation*, p. 30.

Concepts of Space 35

the "location" of the body of Christ in heaven, and the Eucharistic *parousia.*[67]

In criticizing Lutheran and Newtonian receptacle notion of space,[68] Torrance points out the influence of the famous nominalist Occam on Luther: "Luther took over from the Occamists a distinctive notion of 'presence' *(praesentia)* which they developed in their application of Aristotle's concept of place as a containing vessel, according to which a thing may be in a place without being able to be measured according to the space of the place, but without resorting to transubstantiation."[69] Torrance refers to two difficulties with the Lutheran principle: the theories of kenosis and the ubiquity of the body of Christ. On account of the receptacle notion of space, on the one hand, Lutherans had to think of the Incarnation as the self-emptying of Christ into the receptacle of a human body.[70] On account of their concept of space, on the other hand, they had to understand the real presence in the Eucharist as a form of the ubiquity of the body of Christ.

> The receptacle had to be enlarged in order to make it receive the divine nature within its dimensions, and so it was held that the Son of God communicated to the humanity of Christ an infinite capacity enabling it to be filled with the divine full-

---

67 Torrance, *Space, Time and Incarnation*, p. 31. Concerning the "Calvinist extra," Torrance elsewhere explains as follows: "Patristic and Reformed theology have always claimed that the Eternal Logos did enter space and time, not merely as Creator, but as himself made creature, and therefore within the creaturely limits of space and time, and yet did not cease to be what he eternally was in himself, the Creator Word in whom and through whom all things consist and by whom all things derive and continue to have their being. This view was rejected by Lutherans because like the Mediaevals but unlike the Fathers they operated with a receptacle view of space as the place containing within its limits that which occupies it. Hence when Calvin said of Christ that he became man born of the Virgin's womb without leaving heaven or the government of the world, he was interpreted by Lutherans to imply that in the incarnation only part of the Word was contained in the babe of Bethlehem or wrapped in the swaddling clothes in the cradle, and that something was left 'outside' (extra) – hence the so-called 'Calvinist extra'. It was undoubtedly the Lutheran intention to maintain in the fullest possible way real and full incarnation. And that was certainly right" [T. F. Torrance, *Space, Time and Resurrection* (Edinburgh: T&T Clark, 1976), p. 124].
68 Torrance, *Space, Time and Incarnation*, p. 40.
69 Torrance, *Space, Time and Incarnation*, p. 32.
70 Torrance, *Space, Time and Incarnation*, p. 35.

36   *Transcendence and Spatiality of the Triune Creator*

> ness. In regard to the real presence in the Eucharist this took the form of the ubiquity of the body of Christ.[71]

Torrance thinks that the concept of space of Reformed theology is in the same vein with that of space of the church fathers, especially Athanasius, who laid the foundation of the Nicene Creed. Torrance defends this relational, or dynamic, idea of space as being consistent with the space-time relativity of Einstein.

Torrance understands the debate between Reformed and Lutheran theologians over the mode of Christ's presence in the Eucharist as a contest between the receptacle and the relational understandings of space and time. Contrary to the Lutheran view, the Reformed position operated with a relational concept of space much closer to that of the Nicene fathers. However, P. Mark Achtemeier questions the validity of Torrance's conclusion. Even though this analysis is suggestive, its status as a historical claim is open to question, for there is no historical evidence to suggest that the original participants ever thought of this debate as a quarrel between rival means of conceptualizing space. Thus Achtemeier states, "Torrance's argument could perhaps stand credibly as an *ex post facto* logical analysis serving the *constructive* task of reframing the debate in new terms, perhaps in order to assist with ecumenical efforts aimed at overcoming the original impasse."[72]

---

71   Torrance, *Space, Time and Incarnation*, p. 36. Elsewhere he indicates difficulties of the Lutheran theology due to its concept of space as a finite receptacle: "But in actual fact this way of relating the 'immensity' of God to the finite receptacle – *finitum capax infiniti* – usually meant the extension of the human receptacle to contain the divine, e.g. in the doctrine of the ubiquity of the body of Christ, which could hardly avoid a form of monophysitism [...] In Lutheran thought it came to imply a radical disjunction between the divine and the human in which there was no interaction between them" (*Space, Time and Resurrection*, p. 125).

72   P. Mark Achtemeier, "Natural Science and Christian Faith in the Thought of T. F. Torrance," *The Promise of Trinitarian Theology: Theologians in Dialogue with Thomas F. Torrance*, ed. E. Colyer (Lanham, MD: Rowman & Littlefield, 2002), p. 295. D. Bertoloni Meli, "Caroline, Leibniz, and Clarke," *Journal of the History of Ideas* (1999) points out that Leibniz was a Lutheran, and his life-long concern was church union. Meli states that Leibniz's views on substance, gravity, space, and time evolved over several decades hand-in-hand with his theological concerns and were an eminently suitable ground for church reunion (p. 484). This description of Meli is different from Torrance's arguments that Lutheran theology and Newtonian system shared the absolute concept of space and that Reformed and Anglican theology stood much closer to the relational concept of space. In the con-

Concepts of Space

For Torrance, the essential key to the Nicene conception of space is found in the relation of the *homoousion* to the *creation*. He states that "the Lord Jesus Christ, who shares with us our creaturely existence in this world and is of one substance with the Father, is He through whom all things, including space and time, came to be."[73] By incarnation Jesus Christ has come into our spatial realm, although he was not far off before. *Homoousion* means that Jesus Christ is fully present with us in space and time and yet remains present with the Father.[74]

Besides the term *homoousion*, Torrance adopts another term of the Greek fathers to explain the Nicene conception of space. This is the doctrine of the *perichoresis* (περιχώρησις): "The inter-relations of the Father and the Son must be thought out in terms of 'abiding' and 'dwelling' in which each wholly rests in the other."[75] Torrance further explains as follows:

> Creaturely realities are such that they can be divided up in separated places (ἐν μεμερισμένοις) but this is impossible with the uncreated source of all Being, with the Father, Son and Holy Spirit who wholly dwell in each other and who each have room fully for the others in the one God. Now when the Son, who abides in the Father in that way, became incarnate, He became for us the 'place' where the Father is to be known and believed, for He is the τόπος or *locus* where God is to be found.[76]

Then Torrance raises a question: "Christ is 'in' us through sharing our bodily existence, but He is also 'in' the Father through His oneness with Him, but how are we to work out the relation between these two 'in's?" By an analogical account of the relationship which Athanasius offered, Torrance tries to answer this question. He presents Athanasius' concept of *paradeigma* (παράδειγμα): "it [*paradeigma*] is essentially an operational term in which some image, idea or relation is taken from our this-worldly experience to point beyond itself to what is quite new and so to

---

    troversy between Leibniz and Clarke, Meli sees the important role of Caroline, Princess of Wales (ibid.). Pannenberg also refers to this Princess' role in this controversy [Pannenberg, *Toward a Theology of Nature: Essays on Science and Faith* (Louisville: Westminster/John Knox Press, 1993), p. 59].

73   Torrance, *Space, Time and Incarnation*, p. 14.
74   Torrance, *Space, Time and Incarnation*, p. 14.
75   Torrance, *Space, Time and Incarnation*, p. 15.
76   Torrance, *Space, Time and Incarnation*, p. 15f.

38 *Transcendence and Spatiality of the Triune Creator*

help us get some kind of grasp upon it."[77] Thus Torrance proposes the term *paradeigma* should not be translated as "model" or "representation," far less as "archetype," but may be translated as "point." In this situation, Torrance avers, "space is [...] a *differential* concept that is essentially *open-ended*, for it is defined in accordance with the interaction between God and man, eternal and contingent happening." In other words, "space and time are given a sort of trans-worldly aspect in which they are open to the transcendent ground of the order they bear within nature."[78] According to Torrance, this means that the Nicene concept of space is relatively closed on our side, where it has to do with physical existence, but at the same time it is infinitely open on God's side: "Theological statements operate, then, with essentially open concepts – concepts that are relatively closed on our side of their reference through their connection with the space-time structures of our world, but which on God's side are wide open to the infinite objectivity and inexhaustible intelligibility of the divine Being."[79]

Torrance, however, does not fully develop the space concept in relation to God. He writes:

> We do not speak of space-time in relation to God, but we must speak of the 'place' and 'time' of God in terms of his own eternal life and his eternal purpose in the divine love, where he wills his life and love to overflow to us whom he has made to share with him his life and love. 'Time' for God himself can only be defined by the uncreated and creative life of God, and 'place' for God can only be defined by the communion of the Persons in the Divine life.[80]

In this context Torrance refers to the patristic term "perichoresis": "that is why doctrinally we speak of the *'perichoresis'* (from *chora* meaning space or room) or mutual indwelling of the Father, Son and Holy Spirit in the Trinity of God."[81] But Torrance does not set forth his own view on God's spatiality.

Torrance's grappling with theological conceptions of space is an exception to what Moltmann has observed: "Ever since Augustine, there have been many theological meditations on time. But meditations on

77   Torrance, *Space, Time and Incarnation*, p. 16.
78   Torrance, *Space, Time and Incarnation*, p. 18.
79   Torrance, *Space, Time and Incarnation*, p. 20f.
80   Torrance, *Space, Time and Resurrection*, p. 131.
81   Torrance, *Space, Time and Resurrection*, p. 131.

Concepts of Space   39

space are rare."[82] Yet, surprisingly, Moltmann never mentions Torrance in his discussion of space of creation in his *God in Creation*.[83] Unlike Torrance, who has not expanded on the doctrine of perichoresis, Moltmann presents his own view of perichoretic concept of space as a kind of divine space.

6. Perichoretic Concept of Space

According to Moltmann, the metaphor which is associated with the notion of space as a vessel or receptacle is a *feminine* metaphor: "'Mother space', 'the recipient', 'the all-receiving', 'the matrix', are ancient, mythical symbols for the maternal character of Being, in whose great complex and cohesion everything is in safe-keeping."[84] In the Jewish and Christian tradition there are few meditations on space, however, even though there have been many theological meditations on time ever since Augustine.

Just as everything has its time, Moltmann argues, every living thing produces and forms its space. Everything acquires and moulds the environment which belongs to it and is in accordance with its nature. He asserts that *"the ecological concept of space* corresponds to *the kairological concept of time.* Neither time nor space are homogenous."[85] Concerning time and space, Moltmann states, "both are individual, and are created and determined by what happens 'within' them. Without happening they do not exist at all. There is neither an empty time without events, nor an empty space without objects."[86] According to him, if this *ecological* concept of space is employed, one can come nearer to the biblical traditions about creation. Moltmann distinguishes this concept of space from the geometrical concept of space. The objectification of the world through "the modern Cartesian division between an incorporeal

---

82 Moltmann, *God in Creation*, p. 140.
83 Moltmann does mention Torrance twice in his endnotes in different contexts. Cf. Moltmann, *God in Creation*, pp. 326, 347.
84 Moltmann, *God in Creation*, p. 146.
85 *God in Creation*, p. 145.
86 *God in Creation*, p. 145.

40   *Transcendence and Spatiality of the Triune Creator*

mental subject and a world of extended bodies in geometrical space" is alien to the biblical traditions. Thus he maintains that "every living thing has its own world in which to live, a world to which it is adapted and which suits it."[87]

Moltmann also explores the great discussion of theological and scientific problems of space which was carried on between Newton and Leibniz in the seventeenth century. Their controversial point was: "Is absolute space, in which all finite things co-exist, an attribute of the omnipresent God? Or is space nothing more than the extension of objects, and total space therefore the warp and weft of relationships between all conceivable extended things?"[88] Concerning this discussion, Moltmann suggests the following:

> For Leibniz, the real problem was this: if objects did not exist there would be no space either; so if there were no creation there would be neither space nor time. Newton's problem was: if the finite world exists in the eternal space of God, does it not partake of his eternity? And is it not then as eternal as God himself?[89]

For Moltmann this problem can only be solved if we think of creation as a mediation between the relative space of objects and the eternal space of God.[90] He distinguishes three spaces in the doctrine of the world as God's creation: "first, the essential omnipresence of God, or *absolute space*; second, *the space of creation* in the world-presence of God conceded to it; and third, *relative places*, relationships and movements in the created world."[91] Thus, he maintains that "The created world does not exist in 'the absolute space' of the divine Being; it exists in the space God yielded up for it through his creative resolve. The world does not exist in itself. It exists in 'the ceded space' of God's world-presence."[92] Therefore, at the end of his chapter titled "the space of creation," Moltmann concludes, "The space of the world corresponds to God's world-presence, which initiates this space, limits it and interpenetrates it."[93]

---

87   *God in Creation*, p. 147.
88   *God in Creation*, p. 140.
89   *God in Creation*, p. 156.
90   *God in Creation*, p. 156.
91   *God in Creation*, p. 156f.
92   *God in Creation*, p. 156.
93   *God in Creation*, p. 157.

Concepts of Space

Moltmann does not differentiate absolute and relative concepts of space the same way that Einstein and Torrance do. Moltmann distinguishes absolute space as divine space from relative space as the space of creatures. This scheme is a bit confusing. While geometrical space is related to absolute space, ecological space is relative space. In his later work, Moltmann contends for perichoretic space as a divine space rather than as absolute space.

In *God in Creation* Moltmann contends for an ecological concept of space: everything has its space, and every living thing has its living space. In *The Coming of God*, however, he argues that "this ecological concept of space is not sufficient to allow us to apprehend fully the reality of the living. The experience of space is the experience of 'being within another.'"[94] Thus Moltmann says, "Every human person exists in community with other people, and is also for them a living space. Every living thing is as the subject of its own life the object for other life as well. We are inhabitants and inhabited."[95] In other words, we are presence, space and dwelling for one another. He regards this treatment of space as a usage of the *perichoretic concept of space*. This conceptualization reflects a way of apprehending space as mutual in-existence.[96] Perichoresis means *circuminsessio*, mutual indwelling, and mutual *inhabitatio*, "habitation in." This term offers a better way of describing the warp and weft of life than does ecological terms for space, asserts Moltmann. He also asserts that the perichoretic space concept of reciprocal in-existence on the creaturely level corresponds to the concept of the eternal inner-trinitarian indwellings of the divine Persons.[97]

By the scheme of distinction between space concepts as receptacle versus relational, the geometrical concept of space surely belongs to the former. The ecological concept of space seems to belong to the latter; whereas the perichoretic concept of space definitely belongs to the latter. Yet Moltmann does not distinguish the relational concept of space from that of the receptacle. Rather, while supporting the relational concept of space, he presents both an ecological concept of space and a perichoretic concept of space. His treatment of space is applied to his interpretation

---

94 Moltmann, *The Coming of God*, p. 300.
95 *The Coming of God*, p. 301.
96 Cf. *The Coming of God*, p. 298. Here the English translator adds her own footnote: "'In-' here being a preposition of location, not a prefix of negation."
97 *The Coming of God*, p. 301.

42 *Transcendence and Spatiality of the Triune Creator*

of the Godhead. God has this perichoretic space. Yet Moltmann goes on to contend that God is a space. Moltmann employs the idea of *Shekinah* to express God's indwelling in creation. With this term, however, he also seeks to reflect the indwelling of creation within God: "'the space of creation' is its *living space in God*. By withdrawing himself and giving his creation space, God makes himself the living space of those he has created [...] The Creator becomes the *God who can be inhabited*. God as living space of the world is a feminine metaphor, as Plato already observed."[98] Through these statements, Moltmann returns to a receptacle idea of space. He develops this idea by the aid of the Jewish concept of the divine *zimzum*. Thus even though he differentiates three different concepts of space, Moltmann does not distinguish nor favor a relational concept of space over against a receptacle idea of space.

7. Conclusion

This chapter has surveyed the historical development of various conceptualizations of space. Throughout history, we see two rival concepts of space in both physics and theology: receptacle ideas of space and relational ideas of space. The receptacle concept of space has presented difficulties for Christian theology. Even long before Einstein, the patristic fathers seem to have worked with a relational concept of space in developing the Christian doctrines of creation, incarnation, and God's interaction in the world. The infinite receptacle idea of space was developed with Newton's conceptualization of absolute space. In some sense, this concept of space was necessary for the development of the modern natural sciences. However, the relativity theory of Einstein later rejected this notion of absolute space and time. Einstein demonstrated that space and time are not absolute and independent realities unrelated to points of view in this world. They are neither the sense organ of God (Newton) nor *a priori* forms of intuition (Kant). In this regard, Einstein's findings support Augustine's early view that space and time were created with the

---

98   *The Coming of God*, p. 299.

Concepts of Space   43

world; or, Augustine's thought can be seen as an anticipation of the theory of relativity.

Torrance maintains that the patristic fathers employed the relational, not the receptacle concept of space. He further notes consistency between the relational space concept of the patristic fathers and the work of Einstein. Furthermore, Torrance sees the space concept implicit in the work of the Reformed and Anglican theologians as comporting with this relational space concept. Torrance argues that Lutheran theology has assumed a Newtonian view of absolute space. Yet, Torrance's hypothesis is based on implicit, not explicit, aspects of theologies. In his own attempt to speak explicitly about the space of God, Torrance suggests the term *perichoresis*. But he does not fully develop this idea.

Moltmann deals explicitly and in depth with the concept of space. He classifies the concept of space into three concepts: geometrical, ecological, and perichoretic concepts of space. In his doctrine of creation, he proposes an ecological concept of space. But, to present his eschatology, Moltmann deepens his thought on space and presents a perichoretic concept of space. This concept of space is not definitely a receptacle one. The perichoretic concept refers to mutual interpenetration or mutual indwelling. Employing the perichoretic concept of space, Moltmann seeks to present a kind of divine space. In this sense, he goes one step further than Torrance. However, unlike Einstein and Torrance, Moltmann does not clearly distinguish the receptacle idea of space from the relational concept of space. He identifies the omnipresence of God with absolute space and the relationships and movements in the created world with relative space. Absolute space is inadequate to describe the omnipresence of God. With the aid of the concept of the relative space, we can express the divine omnipresence. Thus, this thought is confusing. Nevertheless, Moltmann's work highlights the importance of discussing the relation between God and space, especially if theology is to preserve a doctrine of God as both transcendent and immanent.

Chapter III

Divine Transcendence

1. Introduction

The historical survey presented in Chapter II, above, demonstrated that Einstein's view of space and time, as suggested by his theory of relativity, is quite consistent with the theology of Augustine. In short, both the modern physicist and the ancient theologians would agree that space and time are not absolute but relative in this world. Since the time of Augustine, Christian theologians have given a lot of attention to discussion of time, but not so much to space. Augustine contended that, since God is eternal, he is out of time, or timeless. In opposition to that view, modern theologians have suggested that, to be present, or to interact, in this world of space and time, God must be a kind of a temporal being. Yet, this position threatens the Christian doctrine of divine transcendence, so traditional Christian theology has been reluctant to speak of divine temporality. Modern theologians would have responded that there was widespread influence of Greek philosophy on early Christian theology, especially in the patristic period. So dualism between time and eternity is not a Christian thought, but rather a neo-platonic idea. Thus, the transcendence and temporality of God may not necessarily contradict each other. They may be compatible aspects of the nature and work of God.

This chapter examines the doctrine of divine transcendence in relation to God's spatiality. As with the above suggestion of God's temporality, I will here argue that divine transcendence and spatiality are not incompatible. Yet, instead of asserting a positive statement of God's spatiality, I will set forth a passive, or negative, conclusion that God's transcendence does not necessarily exclude a kind of divine spatiality. In section 2, I will describe the current theological crisis of divine transcendence. Our times are the age of immanence. This emphasis on divine immanence is

46 *Transcendence and Spatiality of the Triune Creator*

desirable as a corrective to historical trends. But to speak of God's immanence without acknowledging God's transcendence is to swing the historical pendulum back too far the other way. In section 3, I will defend the importance of the doctrine of the divine transcendence, over against the view of panentheism, which, as I will maintain, blurs the distinction between God and the world and finally slides into pantheism. In section 4, I will further point out the pitfalls of deism, which holds exclusively to God's transcendence. Deism provides for no discussion of God's interaction with the world. In section 5, attempting to defend theism while also going beyond the apophatic conclusion of Gunton, I will aver that there is some room for speaking of God's spatiality in a way which need not contradict the doctrine of divine transcendence.

2. The Crisis of Transcendence

Concerning Bultmann's demythologizing program as the existential form of a relativized loss of transcendence, Ray S. Anderson asserts that

> One cannot doubt the seriousness with which Rudolf Bultmann wishes to preserve the 'office' of the gospel or to make modern man aware of the real transcendence of God, but in the final analysis, he must be said to have relativized the loss of transcendence to man's self-understanding. This problem of relativizing in this sense could be termed 'solipsism,' which Wittgenstein defines as 'turning all statements into descriptions of my inner life...'"[1]

This relativized loss of transcendence appears in its modern cultural form and can be criticized as presenting a *new paganism*. Yet, paganism cannot be directly equated with atheism. For, as Anderson observes, "paganism is the expression of a strong and vital faith in the presence of the gods."[2] He maintains that "one aspect of a particularly religious form of relativizing is the *worship of transcendence*, even as silence."[3] Finally, Anderson contends, "In this case, religion becomes the face of the silent

---

1   Ray S. Anderson, *Historical Transcendence and the Reality of God: A Christological Critique* (Grand Rapids: Eerdmans, 1975), p. 5.
2   Anderson, *Historical Transcendence and the Reality of God*, p. 5.
3   Anderson, *Historical Transcendence and the Reality of God*, p. 6.

Divine Transcendence

47

gods, but it is also the face of the religious man. This is the ultimate expression of a relativized loss."[4]

Anderson points out two extremes to be avoided in developing a concept of God's transcendence: the "out-and-out" kind of transcendence and the "immanent" kind of transcendence. Anderson explains this as follows:

> The 'out-and-out' kind of transcendence is plain talk for saying that there is an ultimate reality which has its own existence independent of and beyond a contingent and finite world. That is, it is a concept of a fixed point out side of the process by which all moving points take their reckoning. The 'immanent' kind of transcendence, on the other hand, conceives no fixed point external to the contingent world, but transcending relationships within the process of infinite possibility. Immanent transcendence could be likened to the 'transcending' rock just above the head of a mountain climber by which he pulls himself a few higher. When he overtakes (transcends) that point, there is another one always above him. Immanent transcendence includes all of the possibilities of an infinite series of events as well as an infinite self-transcendence of this series.[5]

As examples of immanent transcendence, Anderson points to concepts employed by the process theology of Charles Hartshorne and the ontology of Paul Tillich. Their thoughts collapse into immanent transcendence. "[L]ike process theology," Anderson says, "the ontology of Tillich appears to be no more than a regulative 'symbol' of ultimate reality to give man the 'courage to be.' Without the regulative function of the symbol, Tillich's language of transcendence collapses into immanent transcendence."[6]

On the other hand, if the "out-and-out" kind of transcendence is the case with God, then we cannot know God nor speak positively about God at all. A corollary of this thought is so-called "negative theology," or a "[w]ay of knowing God through negation and silence, indicating his transcendence of all thought and expression."[7] Since an "out-and out" transcendent God is beyond all being, "we cannot say anything about what God is, but can only say something about what he is not." Citing

---

4   Anderson, *Historical Transcendence and the Reality of God*, p. 7.
5   Anderson, *Historical Transcendence and the Reality of God*, p. 24f.
6   Anderson, *Historical Transcendence and the Reality of God*, p. 29.
7   David F. Ford, ed., *The Modern Theologians: An Introduction to Christian Theology in the Twentieth Century* (Oxford: Blackwell Publishers, 1997, Second Edition), p. 748.

48   *Transcendence and Spatiality of the Triune Creator*

Gregory Nazianzen, however, Torrance disputes negative theology: "if we cannot say anything positive about what God is, we really cannot say anything accurate about what he is not."[8] According to Torrance, the Nicene doctrine of God clearly rejects "any conception of God so utterly transcendent that he cannot accommodate himself to natures other than his own, and does not interact with the world."[9] Epistemologically, therefore, Torrance states, "what the doctrine of the Holy Trinity is about, and what the Nicene Creed is about: that we may really know God the Father Almighty in a positive way."[10]

For Torrance, we human beings can really know God as He is in Himself, due to God's own condescension. "If even the holy cherubim or seraphim cover their faces in the immediate presence of God," he asks, "how much more must human beings, for how can they look upon God, really know God as he is in himself, and live?" Then Torrance immediately answers, "Yet that is precisely what the astonishing condescension (συγκατάβασις) of God in Jesus Christ has made possible, accurate but devout knowledge of God in accordance with his divine nature."[11] However, Torrance also acknowledges the necessary limits of our knowledge of God: "We cannot embrace all that God is within our knowing of him, for he breaks through all our finite conceptions and definitions."[12] Nevertheless, Torrance asserts, God as an object of our knowledge is "not a part of God that we know, for God cannot be divided up in parts; rather do we know in part the whole God who exceeds what we can apprehend within the embrace of our human thought and speech."[13]

---

8   T. F. Torrance, *The Trinitarian Faith: The Evangelical Theology of the Ancient Catholic Church* (Edinburgh: T&T Clark, 1995), p. 50.

9   Torrance, *The Trinitarian Faith*, p. 65.

10   Torrance, *The Trinitarian Faith*, p. 67.

11   Torrance, *The Trinitarian Faith*, p. 67f.

12   Torrance, *The Trinitarian Faith*, p. 53. See also Torrance, *Theological Science* (Edinburgh: T&T Clark, 1996), p. 344, where he writes, "It is God who has given and revealed Himself in Israel and in the Incarnation for men to know and understand, but He remains the Lord God transcendent in His eternal and infinite Being who cannot be comprehended even when we apprehend Him [...] God has imparted Himself to us within our human existence in space and time, within historical being on earth, and yet as such He remains the Lord God, Father, Son and Holy Spirit, who cannot be resolved into the forms and modes of His self-impartation."

13   Torrance, *The Trinitarian Faith*, p. 53.

Divine Transcendence

Torrance states that we find and know God in Jesus Christ where He "has condescended to reveal Himself to us within our creaturely existence and contingency, and has assumed our humanity to meet us as man to man and to make Himself known to us within the conditions and limitations of our earthly life, within our visible, tangible, temporal flesh."[14] The incarnation of God the Son, Torrance states, revealed that "God is neither some utterly unknowable transcendent deity nor some pantheistic force immanent in the unchangeable laws of natural process."[15] We cannot understand the transcendent God as the unmoved mover in the Aristotelian sense: "God does not hold himself aloof from the world like some unmoved mover who keeps his power to himself, but moves out of himself to become one with his creatures, while remaining what he is in all his eternal constancy as God, in such a way as to share ungrudgingly with them the power of his divine life and love in what Christians learned to call *grace*."[16] Torrance asserts, by his very nature, even in his condescension, God transcends the range of all human knowledge and speech.[17] Referring to a saying of Athanasius that "Thus far human knowledge goes. Here the cherubim spread the covering of their wings," to use an Old Testament analogy, Torrance suggests, "God reveals him-

---

14   Torrance, *Theological Science*, p. 46. In several places, Torrance refers to the condescension of God. See *Theological Science*, p. 51f, 190 and *The Trinitarian Faith*, p. 31f. Elsewhere, quoting Athanasius, Torrance comments on the divine condescension: "The Saviour came, Athanasius repeatedly insisted, not for his own sake but for our sake and for our salvation. The concepts of 'the economy of the incarnation,' …'the economic condescension'… of God the Son, or of 'the advent in the flesh,' … 'the divine and loving condescension and becoming man'… etc., that is, of God's loving assumption of our actual nature and condition in space and time, all for our sake, were quite dominant in his thought" (*The Trinitarian Faith*, p. 148).

15   T. F. Torrance, *Divine and Contingent Order* (Edinburgh: T&T Clark, 1998), p. 65.

16   Torrance, *Divine and Contingent Order*, p. 65.

17   Cf. Torrance, *The Christian Doctrine of God: One Being Three Persons* (Edinburgh: T&T Clark, 2001), p. 77, writes that "even while God condescends to make himself known within the lowly conditions of human existence in space and time, as he has done in the incarnation of his Word and Truth in Jesus Christ, nevertheless by his very nature he transcends the range of all human knowledge and speech." Similarly, Torrance states that "God reveals himself to us in such a way as still to be veiled from us in the infinite depths of his ultimate Being, for he does not surrender his transcendence or sovereign freedom but remains the absolute Lord of what he reveals and of our knowing of him" (ibid., p. 81).

50 *Transcendence and Spatiality of the Triune Creator*

self to us only under the cover of his hand, and in such a way that we are unable to get behind the back of his revelation to see him face to face in his unveiled Glory."[18]

Gordon Kaufman proposes an interpersonal model for describing God's transcendence.[19] By this interpersonal transcendence, he tries to avoid an arbitrary opposition between our world and the "beyond." Concerning Kaufman's view, Thomas F. Tracy observes that Kaufman's strategy is to locate some point within our experience at which the concept of transcendence already has a meaning, so that this experienced transcendence can serve as a model for our relation to that which transcends us ultimately. Kaufman argues that "there are important respects in which a human self is beyond the reach of others' knowledge unless he chooses to make himself known."[20] And, such is the case with God as well.

Anderson takes a position similar to Kaufmann's interpersonal model of God's transcendence. Anderson holds that "[t]o speak of transcendence [...] in the mode of the personal, is to speak of a limit which is constituted by spirit itself which must be crossed over."[21] Anderson warns the danger of the "out-and-out" kind of transcendence: "When we start from the infinity of God and attempt to reach his particular reality, we destroy the decisive character of the encounters of God in a concrete sense and end up with the theological ambivalence which is characteristic of the silence of the gods."[22] Then Anderson identifies God's self-revelation as His transcendence. Anderson argues that "God's temporality, his self-revelation, is his transcendence [...] God's transcendence is his act of concretizing himself as Spirit with and in human existence."[23] Yet, here it seems that Anderson's so-called "historical transcendence" collapses into an "immanent" kind of transcendence as does Kaufman's

---

18 Torrance, *The Christian Doctrine of God*, p. 81.

19 Cf. Gordon D. Kaufman, *God the Problem* (Cambridge, MA: Harvard University Press, 1972), "Transcendence without Mythology," pp. 41–71 and "Two Models of Transcendence," pp. 72–81. In the latter article, he contrasts his interpersonal transcendence model with that of teleological transcendence. Kaufman presents two representatives of the latter position as Paul Tillich and Martin Heidegger.

20 Thomas F. Tracy, *God, Action, and Embodiment* (Grand Rapids: Eerdmans, 1984), p. 63.

21 Anderson, *Historical Transcendence and the Reality of God*, p. 18.

22 Anderson, *Historical Transcendence and the Reality of God*, p. 112.

23 Anderson, *Historical Transcendence and the Reality of God*, p. 122f.

Divine Transcendence 51

interpersonal model. We reveal ourselves through our words and acts. In this sense we transcend ourselves. Anderson and Kaufman understand God's transcendence as His self-revelation. But, this kind of transcendence falls short of that of the triune Creator. It is just a self-transcendence.[24]

Anderson admits that the term of "historical transcendence" is a somewhat odd combination of words. Some might say that it is an apparent confusion of language games in a Wittgensteinian sense. Yet, Anderson defends the formulation of this term as being "based on the thought of Ronald Gregor Smith, who may well have been the first to recognize and attempt to develop Dietrich Bonhoeffer's fragmentary remarks concerning transcendence and worldliness into a specific theology of transcendence as historical experience."[25]

Most contemporary theologians agree with the contention that classical theism unduly emphasized the transcendence of God. The divine transcendence of classical theism betrays a tendency to the "out-and-out" kind of transcendence. As a reaction to this partiality, many scholars have more recently asserted the importance of divine immanence. This reactive emphasis presents theology today with a crisis concerning God's transcendence.

---

24 Peter D. Fehlner, *Theological Studies* 37.4 (Dec. 1976), p. 689, evaluates Anderson's study as sympathetic to process theology at many points. Regarding the doctrine of the divine transcendence of process theology, John Cobb, Jr. and David Griffin themselves present the following critical literatures: "Langdon Gilkey, in *Maker of Heaven and earth: A Study of the Christian Doctrine of Creation* (Doubleday & Company, Inc., 1959), argued that a more transcendent deity is required than that of Whitehead. More recently, Robert C. Neville, in *God the Creator: On the Transcendence and Presence of God* (Chicago: The University of Chicago Press, 1968), has reemphasized this criticism in careful analyses of Whitehead and Hartshorne" [John B. Cobb, Jr. and David Ray Griffin, *Process Theology: An Introductory Exposition* (Philadelphia: The Westminster Press, 1976), p. 184].

25 Anderson, *Historical Transcendence and the Reality of God*, p. xiv.

52 *Transcendence and Spatiality of the Triune Creator*

3. Transcendence versus Panentheism

Christians worship a God who is a transcendent being over the world of creation. Of course this is not a unique characteristic of Christianity, but is an aspect of Judaism and Islam as well. If God is not transcendent beyond the universe, then God is no longer the one God in every sense. Throughout church history, orthodox theologies, especially classical theism, have emphasized this transcendent aspect of God. God is infinite, while other creatures are finite. God is eternal, whereas creatures are temporal. God is in heaven, while human beings are on earth. As a reaction to an often one-sided emphasis on this divine transcendence, contemporary theologians have tended toward panentheism over theism. Panentheists aver that their panentheism is the appropriate alternative to either theism or pantheism.

Process theologians contend that God is dependent upon the world, while also this world is dependent upon him. By contrast, so-called "classical theism" asserts that God by nature is independent of this world, even though this world of creatures is absolutely dependent upon God the creator, and God is active in relation to the world. The panentheism of process theology denies the doctrine of creation out of nothing. For "the doctrine of creation means that while the world is dependent upon God, God is not dependent on the world."[26] Thus the relationship between God and the world in the panentheism of process theology is not unilaterally but mutually dependent.

Stanley Grenz and Roger Olson indicate that Hegel and Schleiermacher are panentheists. For Hegel, God cannot be God without this world. God is not a self-sufficient being. Rather God needs this world for his self-actualization. The world history is also a history of God. In opposition to traditional theism in which God is entirely self-sufficient in relation to the world, Hegel's panentheistic approach represents God and the world as inseparable yet distinct realities. Thus, for Grenz and Olson, the system of Hegel is the "work of a radical immanentist."[27] For Schleiermacher, it is not clear whether God exists independently from

---

26   Ronald Nash, "Process Theology and Classical Theism," in *Process Theology* (Grand Rapids: Baker, 1987), p. 15.
27   S. Grenz and R. Olson, *20th-Century Theology*, p. 39.

Divine Transcendence

this world. Schleiermacher tries to correlate God and this world and make them inseparable. According to Grenz and Olson, all panentheistic vision about the relationship between God and the world raises serious problems for the doctrine of grace.

> How can God's redemption of the world be gracious if it is not absolutely free? In tension with his emphasis on God's absoluteness, the ultimate upshot of which is the unavoidable loss of personal relationship. A true relationship with God becomes impossible because it would involve reciprocity of action and response.[28]

Thus, Grenz and Olson object to Schleiermacher's panentheistic vision of the God-world relationship.

In contrast to pantheism, classical theism holds that God's essential nature and actual existence are coextensive or identical. God is immutable, or changeless, because God is perfect. He cannot become more perfect. Thus there is no potentiality in the divine nature or within God's existence that becomes actualized in response to the world. God's knowledge, power, goodness, perfection, infinity, threeness in one, and other attributes are all everlastingly fully actual and therefore changeless even in relation to the world. God is active in the world, but this activity does not change God himself.

At this time, process theology does not monopolize panentheism.[29] Feminist theology and some liberationists assume a panentheistic view.[30] In fact, the list of panentheists includes such notable figures as Jürgen Moltmann,[31] Wolfhart Pannenberg,[32] Philip Clayton,[33] Arthur Peacocke,[34] and Ian Barbour.[35]

---

28   S. Grenz and R. Olson, *20th-Century Theology*, p. 50.
29   Nash, *The Concept of God*, p. 23, identifies the major proponents of process theology as follows: Charles Hartshorne, Schubert Ogden, John B. Cobb Jr., Norman Pittenger, Daniel Day Williams, Bernard Loomer, H. N. Wiemann, David Ray Griffin, Peter Hamilton, and Ewert Couisins.
30   See Elizabeth A. Johnson, *SHE WHO IS: The Mystery of God in Feminist Theological Discourse* (New York: Crossroad, 1995) and Sallie McFague, *The Body of God: An Ecological Theology* (Minneapolis: Fortress Press, 1993). See also Leonardo Boff, *Ecology and Liberation: A New Paradigm* (Ecology and Justice), trans. John Cumming (Maryknoll, NY: Orbis Books, 1995).
31   Moltmann's version of panentheism can be dubbed as Christian panentheism or eschatological panentheism. See Moltmann, *God in Creation* (Minneapolis: Fortress Press, 1993). In her recent Ph. D. dissertation, Oksu Shin presents Moltmann's panentheism as "Trinitarian, voluntary, and eschatological panenthe-

54   *Transcendence and Spatiality of the Triune Creator*

Panentheists commonly complain that classical theism unduly emphasizes the transcendence of God, but that pantheism unduly emphasizes His immanence. They contend that panentheism avoids these two extremes. However, many of these arguments are flawed by an over-identification of classical theism with deism; in other words, theologians may tend to buy into panentheism because of a misinterpretation of the traditional view of divine transcendence.

Within this discussion, John Polkinghorne has a peculiar position. Unlike his fellow so-called "scientists as theologians," such as Peacocke and Barbour, he opposes panentheism. Polkinghorne points out an undue emphasis of classical theism on divine transcendence. Nevertheless, he does not think of panentheism as an apt solution to Christian theology. Polkinghorne writes: "Many of us would share a recognition of the need to correct classical theism's undue emphasis on the transcendent remoteness of God, without feeling that this implied a necessity to adopt panentheistic language. It simply requires a recovery of the balancing orthodox concept of divine immanence."[36]

---

ism" [Oksu Shin, "The Panentheistic Vision of the Theology of Jürgen Moltmann" (Fuller Theological Seminary 2002 Ph. D. Dissertation)].

32   Concerning whether or not Pannenberg is a panentheist, there is a controversy. The co-writers of *20th-Century Theology* split into opposite positions on this issue. Stanley Grenz does not agree with the position of his colleague Roger Olson who in his dissertation contends that Pannenberg is a panentheist [Stanley Grenz, *Reason for Hope: The Systematic Theology of Wolfhart Pannenberg* (Oxford University Press, 1990)]. Olson presents Pannenberg's version as eschatological panentheism. See "Trinity and Eschatology: The Historical Being of God in the Theology of Wolfhart Pannenberg" (Rice University 1984 Ph. D. Dissertation). In my Th. M. thesis I dub Pannenberg's as 'voluntary panentheism'. See "The Voluntary Panentheism of Wolfhart Pannenberg" (Calvin Theological Seminary 1999 Th. M. Thesis).

33   Philip Clayton, *God and Contemporary Science* (Grand Rapids: Eerdmans, 1997).

34   A. Peacocke, *Theology for a Scientific Age: Being and Becoming – Natural, Divine, and Human* (Minneapolis: Fortress Press, 1993, enlarged ed.).

35   I. Barbour, *Religion and Science: Historical and Contemporary Issues* (San Francisco: Harper Collins, 1997).

36   J. Polkinghorne, *Scientists as Theologians: A Comparison of the Writings of Ian Barbour, Arthur Peacocke and John Polkinghorne* (London: SPCK, 1996), p. 33. Polkinghorne, however, accepts panentheism as the final state of eschaton. Thus, though he opposes panentheism, Polkinghorne is in favor of Moltmann's idea of divine *zimzum* [J. Polkinghorne, *Science and Providence: God's Interaction with the World* (London: SPCK1989), p. 22]. Concerning the advocacy of classical the-

Divine Transcendence 55

Panentheism holds the position that God includes this world but is not included by it. Thus, most panentheists deny the doctrine of the creation out of nothing. For them, creation is not out of nothing but within God in some sense. In this scheme, God is not totally detached from the world. God and the world coexist. In this case the reality of the world itself is damaged. Colin Gunton criticizes this aspect of panentheism:

> [A] distinction between God and the world is necessary not only to preserve the autonomy of God's action – its character as authentically divine action – but also for the sake of the world. If the world is too closely tied to the being of God, its own proper reality is endangered, for it is too easily swallowed up into the being of God, and so deprived of its own proper existence.[37]

Gunton thus asserts that "Panentheism cannot finally be distinguished from pantheism, because it does not allow the other space to be itself."[38]

In the same vein, criticizing the dipolar God of process thought as a hoax or mirage, Bruce Demarest charges: "The primordial pole of God, which possesses no actuality, in fact possesses no reality. The attempt to impose upon the system a transcendent, timeless anchor in the form of the primordial nature smacks of a desperate attempt to forestall the systems' collapse into a radical immanentism and pantheism."[39] For Demarest, panentheism is not a brilliant stroke of genius but a desperate act of expediency. Panentheism does have nothing to prevent its inevitable collapse into sheer pantheism. Karl Barth also opposes both of pantheism and panentheism. He contends that panentheism is in a worse state than pantheism. Since pantheism and panentheism do not distinguish God and the world definitely, Barth thinks, they do threaten God's freedom or absoluteness.[40]

---

ism, see Richard A. Muller, "Incarnation, Immutability, and the Case for Classical Theism," in *Westminster Theological Journal* 45 (1983).

37 C. E. Gunton, *The Triune Creator: A Historical and Systematic Study* (Grand Rapids: Eerdmans, 1998), p. 65f.

38 Gunton, *The Triune Creator*, p. 142.

39 Bruce Demarest, "Process Trinitarianism," *Perspectives on Evangelical Theology*, ed., Kenneth S. Kantzer and Stanley N. Gundry (Grand Rapids: Baker, 1979), p. 35.

40 K. Barth, *Church Dogmatics* II/1 (Edinburgh: T&T Clark, 1957), p. 312, states, "The mythology of a merely partial and to some extent selected identity of God with the world, which under the name of panentheism has been regarded as a better

56 *Transcendence and Spatiality of the Triune Creator*

Even though panentheists think that their position incorporates both the transcendence and immanence of God more effectively than does classical theism, panentheism's explanation of divine transcendence is inadequate, as Peter Kreeft explains:

> God cannot be a part of the universe. If he were, he would be limited by other parts of it. But God is the *Creator* of all things, giving them their total being. He cannot be one of them, or the totality of them – for each one of them, and so the totality of them, must be given being, must receive being from God. So God must be *other* than his creation. This is what we mean by the *transcendence* of God.[41]

Despite panentheism's appealing interest in the immanence of God, it is not an adequate alternative to theism, for panentheism too easily loses sight of the divine transcendence which is essential to the Christian doctrine of creation out of nothing.

Perhaps one problem is that panentheists have tended to identify traditional or classical theism with deism. They have thus sought a balance between deism, which exclusively stresses God's transcendence, and pantheism, which exclusively stresses divine immanence. Yet, it is possible to find both sides already presented in classical theism, although perhaps in an unbalanced way. Therefore, John Polkinghorne contends that we simply require some revision of orthodox concepts of divine transcendence and immanence.[42] Though we might suggest that theism needs some revision, we need not consent to the view that panentheism is a viable alternative proposal to traditional or classical theism.

In his book, *God, Creation, and Contemporary Physics*, Mark W. Worthing offers a valuable discussion on the fruitfulness of dialogue between theology and science. Concerning the importance of the transcendence of God, he says, "If God is not transcendent (completely other) from the universe, then we would have to say that God, inasmuch as God would be contained within the closed system of our physical

---

possibility than undiluted pantheism, is really in a worse case than is that of the latter."

41 P. Kreeft and R. K. Tacelli, *Handbook of Christian Apologetics* (Downers Grove, IL: InterVarsity, 1994), p. 93.

42 Polkinghorne, *Scientists as Theologians*, p. 33 f. Elsewhere Polkinghorne says: "There are distinctions between God and the world that Christian theology cannot afford to blur" [Polkinghorne, *Science and Providence: God's Interaction with the World* (London: SPCK, 1989), p. 16].

Divine Transcendence 57

universe, is also subject to the increasing-entropy principle of the second law."[43] In the dialogue with science, theology must not forget the divine transcendence. For Worthing, any theology that does not recognize the transcendence of God would have a difficulty explaining how God could avoid necessarily "winding down" along with the universe. Worthing furthermore points up the problem of how a nontranscendent God could in any real sense be considered the creator of space and time.

Worthing likewise emphasizes the importance of the doctrine of the creation out of nothing for the transcendence of God. In order to support the transcendence of God, the doctrine of creation can not be held as an extension of his being but rather must be seen as an act of God, based in his very being. As Gunton argues, creation is an unnecessitated act in which something external to and ontologically other than that being is created.[44] Worthing asks: "Without the traditional doctrine of the transcendence of God, modern physics, as it moves closer and closer to an initial singularity, t=0, or a *creatio ex nihilo*, brings the existence of God increasingly into question. If God does not transcend the physical universe, what place is left for God within it?"[45] Two possible alternatives may be considered: a God of the gaps which will ultimately remain; or a crass pantheism in which we rename the physical universe "God." Neither of these options is particularly attractive to Christian theology. Thus, Worthing argues, "we are left with the creation *ex nihilo*, which alone does justice to the Christian understanding of God as Creator."[46]

Dietrich Bonhoeffer expressed the shortcomings of the "God of the gaps" in one of his later prison letters. He writes: "It has again brought home to me quite clearly how wrong it is to use God as a stop-gap for the incompleteness of our knowledge. If in fact the frontiers of knowledge are being pushed further and further back (and that is bound to be the case), then God is being pushed back with them, and is therefore continually in retreat. We are to find God in what we know, not in what we don't know."[47] This "God of the gaps" has not fared well in modern

---

43   M. W. Worthing, *God, Creation, and Contemporary Physics* (Minneapolis: Fortress Press, 1996), p. 84.

44   Gunton, *The Triune Creator*, p. 66.

45   Worthing, *God, Creation, and Contemporary Physics*, p. 108.

46   Worthing, *God, Creation, and Contemporary Physics*, p. 108.

47   Dietrich Bonhoeffer, *Letters and Papers from Prison*, Eberhard Bethge, ed., (New York: Macmillan, 1979, Enlarged Ed.), p. 311.

58   *Transcendence and Spatiality of the Triune Creator*

theology. John Polkinghorne explains the "God of the gaps" and presents his objection to this concept of God: "The invocation of God as an explanation of last resort to deal with questions of current (often scientific) ignorance. ('Only God can bring life out of inanimate matter,' etc.). Such a notion of deity is theologically inadequate, not least because it is subject to continual decay with the advance of knowledge."[48] Arthur Peacocke asserts that divine interventionist view is related to the concept of "God of gaps." So he maintains that "it would be extremely unwise for any proponent of theism to attempt to find any gaps to be closed by the intervention of some nonnatural agent, such as a god."[49] For "as these gaps were filled by new knowledge," according to Peacocke, "'God' as an explanation became otiose."[50] But now we have "'a God of the *permanent* and (to us) *unclosable* gaps' [...] 'within' the flexibility we find in these (to us) unpredictable situations in a way that could never be detected by us."[51] Thus, Peacocke contends that, "there would then be no possibility of such a God being squeezed out by increases in scientific knowledge."[52] As in Peacocke, Nancey Murphy opposes both interventionist accounts of divine action and "God of gaps" accounts of divine action. Interventionism supposes that God should violate the laws of nature he has established. It seems unreasonable. The objection to the view of "God of gaps" is due to epistemological reasons – "science will

---

48   Polkinghorne, *The Faith of a Physicist: Reflections of a Bottom-Up Thinker* (Minneapolis: Fortress Press, 1996), p. 197.
49   A. Peacocke, *Intimations of Reality*, p. 57. Elsewhere Peacocke contends as follows: "That God is not to be conceived of as a 'God of the gaps' in our knowledge of the natural, including the human, world is widely agreed – for such a 'god' has the habit of shrinking to nothingness as our knowledge increases" (Peacocke, *Theology for a Scientific Age*, p. 144).
50   Peacocke, "God's Interaction with the World: The Implications of Deterministic 'Chaos' and of Interconnected and Interdependent Complexity," in R. J. Russell, N. Murphy and A. R. Peacocke, eds, *Chaos and Complexity: Scientific Perspectives on Divine Action* (Vatican City State: Vatican Observatory Publications, 1997, 2nd Ed.), p. 277.
51   Peacocke, "God's Interaction with the World," p. 277.
52   Peacocke, "God's Interaction with the World," p. 277.

Divine Transcendence 59

progress and close the gaps."[53] This "God of the gaps" is not the transcendent Creator of the world.[54]

Worthing rightly observes that "A God who is not transcendent is not capable, even in principle, of miraculous intervention because such a God would be necessarily subject to and limited by the laws of nature."[55] In the summary portion of his chapter which deals with the continuous activity of God in the universe, Worthing concludes,

> The dialogue with physics, even in the fields most closely concerned with questions of God's immanence, seems once again to point to the importance of confessing God's transcendence. If God were only immanent, not only would God be confined by physical law, but advances in physics would seem on the verge of making God homeless.[56]

While we seek to describe and celebrate the divinely immanent activity in the universe, we still cannot lose sight of the transcendence of God as the creator. Otherwise, we dispel God from our universe.

As stated above, undue emphasis of classical theism on God's transcendence has the weakness of providing a poor explanation for present divine activity in the universe. Along with the divine transcendence, therefore, we should hold the divine immanence. The God of deism is only transcendent not immanent. God is in the upper heaven, but does not have a place in this world. Nevertheless, this God is not the God of the Bible nor of classical theism. In the Bible we find distinct vestiges of God's continuous activities in the world, including human history, as well God's Lordship over the world.

---

53 N. Murphy, "Divine Action in the Natural Order," in R. J. Russell, N. Murphy and A. R. Peacocke, eds, *Chaos and Complexity: Scientific Perspectives on Divine Action*, p. 343.

54 Unlike Polkinghorne, Peacocke, and Murphy, however, Thomas F. Tracy contends that "God would be the creator not only of natural law but also of the indeterministic gaps through which the world remains open to possibilities not exhaustively specified by its past" (T. F. Tracy, "Particular Providence and the God of the Gaps," in R. J. Russell, N. Murphy and A. R. Peacocke, eds, *Chaos and Complexity: Scientific Perspectives on Divine Action*, p. 294). Tracy uses the term "gaps" in a neutral sense. While I oppose the view of the God of the gaps, I will contend for the gaps for God's action in the world in this book. See the section on Agency of Chapter VI.

55 Worthing, *God, Creation, and Contemporary Physics*, p. 157.

56 Worthing, *God, Creation, and Contemporary Physics*, p. 158.

60   *Transcendence and Spatiality of the Triune Creator*

In this study, I adopt a relational notion of divine transcendence and immanence. In their introduction of the book *20th-Century Theology*, Stanley Grenz and Roger Olson contend that, like divine immanence, divine transcendence is a mode of God's relation to the world including human beings. They explain:

> At its best Christian theology has always sought a balance between the twin biblical truths of the divine transcendence and the divine immanence. On the one hand, God relates to the world as the Transcendent One. That is, God is self-sufficient apart from the world. God is above the universe and comes to the world from beyond [...] On the other hand, God also relates to the world as the Immanent One. This means that God is present to creation. The divine one is active within the universe, involved with the processes of the world and of human history.[57]

Grenz further develops this idea of the relational God in his later book, *Theology for the Community of God*. He holds that both transcendence and immanence describe two foundational aspects of God's relationship to creation. "In conceiving of God," Grenz claims, "we dare neither place him so far beyond the world that he cannot enter into relationship with his creatures nor collapse him so thoroughly into the world processes that he cannot stand over the creation which he made."[58]

Grenz and Olson have summarized modern theological history with reference to the twin biblical truths of the divine transcendence and the divine immanence. "Where such balance is lacking," they say, "serious theological problems readily emerge." As Grenz and Olson observe, "an overemphasis on transcendence can lead to a theology that is irrelevant to the cultural context in which it seeks to speak, whereas an overemphasis on immanence can produce a theology held captive to a specific culture."[59] The main issue to maintain a balanced theology of God is to coherently explain these twin truths, which in many instances seem to be contradictory to each other. Grenz and Olson seek to hold both doctrines of them in a "creative tension." Thus Grenz says that God is "that reality who is present and active within the world process. Yet he is not simply

---

57   S. J. Grenz and R. E. Olson, *20th-Century Theology: God and the World in a Transitional Age* (Downers Grove: InterVarsity Press, 1992), p. 11.

58   S. J. Grenz, *Theology for the Community of God* (Grand Rapids: Eerdmans, 1994), p. 81.

59   Grenz and Olson, *20th-Century Theology*, p. 12.

Divine Transcendence 61

to be equated with it, for he is at the same time self-sufficient and 'beyond' the universe."[60]

4. Immanence versus Deism

The split between transcendence and immanence took place with the emergence of natural sciences in modern time. At this time, the deistic response to God's being pushed out of the world was to suggest that "God has no further causal influence on the world beyond creation, because of the difficulty of conceiving actual incursions ('interventions') of God into the world."[61] Pannenberg regards deism as the consequence of the introduction of the principle of inertia in modern physics.[62] Therefore, among his five theological questions to scientists, the first and most fundamental one he puts concerns the principle of inertia and the divine conservation.[63]

The seed of deism can be found in Descartes's dualism between mind and matter. Theologically, according to Gunton, the long-term effect of this dualism of Descartes was disastrous, but in the short term the dualism was encouraged by a theology which taught the scientist to take the created universe seriously in itself, and not as interesting only as pointing beyond itself to another eternal world. Descartes's dualism, which

---

60 Grenz, *Theology for the Community of God*, p. 81.
61 Philip Clayton, *God and Contemporary Science* (Grand Rapids: Eerdmans, 1997), p. 189.
62 Pannenberg, *Toward a Theology of Nature*, p. 35.
63 Pannenberg, *Toward a Theology of Nature*, p. 17f. Here are his five questions to scientists: 1. Is it conceivable, in view of the importance of contingency in natural processes, to revise the princip[le] of inertia or at least its interpretation? 2. Is the reality of nature to be understood as contingent, and are natural processes to be understood as irreversible? 3. Is there any equivalent in modern biology of the biblical notion of the divine spirit as the origin of life that transcends the limits of the organism? 4. Is there any positive relation conceivable of the concept of eternity to the spatiotemporal structure of the physical universe? 5. Is the Christian affirmation of an imminent end of this world that in some way invades the present somehow reconcilable with scientific extrapolations of the continuing existence of the universe for at least several billions of years ahead?

62 *Transcendence and Spatiality of the Triune Creator*

cut mind off from matter, finally cut God from the universe.[64] Descartes did not take inertia simply as a manifestation of a force of perseverance within the body itself. Rather, he took the principle of inertia as manifesting the immutability of God who preserves his creatures in the same form in which He created them. However, the same principle of divine immutability prevented Descartes from ascribing to God the changes that occur in the world of creation. Thus Pannenberg contends, "When the assumption that movement is intrinsic to the bodies themselves was combined with the principle of inertia, the need for the cooperation of God as first cause became superfluous in the explanation of natural processes."[65]

Isaac Newton rejected Descartes's reduction of movement to the concept of body. Newton replaced it with a conception of force as a field that may impress movements upon bodies even over great distances in space. Unlike Descartes, Newton took inertia as a manifestation of a force of perseverance within a body. Newton sought to avoid a mechanical conclusion to the explanation of natural processes apart from God. Nevertheless, God seems to have been rendered superfluous in the system of Newton. Pannenberg describes this problem of Newton as follows:

> But at this point, with his general conception of force, Newton was not successful, at least not in the judgment of his own age. Instead, the combination of Newton's interpretation of inertia in terms of a force that is inherent in bodies with the reduction of force to a body and to its mass contributed in a decisive way in the course of the eighteenth century to the removal of God from the explanation of nature.[66]

---

64   C. E. Gunton, *The Triune Creator*, p. 126f.

65   Pannenberg, *Toward a Theology of Nature*, p. 31. Polkinghorne criticizes Pannenberg's notion of inertia as "a rather old-fashioned concept of inertia." According to Polkinghorne, "Pannenberg seems to treat as if it were a kind of burdensome constraint on the freedom of creation, or even of the Creator." Polkinghorne writes: "While modern physics gives an important role to conservation laws, there seems no need to make them carry such a load of metaphysical freight. They are consequences of the symmetries of creation and can easily be understood as expressions of the Creator's will rather than impositions on it. Fruitful physical process requires a degree of stability as well as a degree of flexibility" [Polkinghorne, *Belief in God in an Age of Science* (New Haven: Yale University Press, 1998), p. 82].

66   Pannenberg, *Toward a Theology of Nature*, p. 31.

Divine Transcendence

63

However, the direct charge that Newton was a deist is not appropriate. Pannenberg argues that that charge is oversimplified. Due to good theological reasons, Newton resisted the direction in which the mechanical world-view appeared to be leading. Nevertheless, Newton inadvertently opened the door to deism and mechanism.[67] Gunton asserts that this effect resulted from Newton's conceptions of absolute and relative space and time, which are deeply problematic:

> They [absolute space and time] introduce, by the back door, so to speak, another form of the very dualism that had marked the earlier Babylonian captivity of the doctrine of creation. The idea that there is, behind the world as we experience it, another 'absolute' world, introduces a breach in our experience. The world we experience and experiment with is not the finally real world, which is something there but unknown and unknowable. Einstein was later to see that this was the Achilles' heel of Newton's philosophy.[68]

In his early Christology, Gunton argues that, since absolute space is externally related to God, to speak of God's co-presence with a particular spatial being becomes problematic. Gunton proposes that space is not to be seen in terms of exclusiveness (absolute space) but of interpenetration *(perichoresis)*. Then, Gunton claims, we can conceive of God at once as Creator of space and as able to relate himself internally to parts of it. Deism can not account for God's internal relation to the world. For, as Gunton states, "deism is a modern expression of the absoluteness of space and the external relatedness of God to it."[69]

For Kant, space and time are absolute in one sense, as conditions for any experience of the phenomenal world. But they seem to be subjective in the sense that they are conditions supplied by our minds to make ex-

---

67   Cf. Edward B. Davis, "Newton's Rejection of the 'Newtonian World View': The Role of Divine Will in Newton's Natural Philosophy," *Fides et Historia* 22 (Sum 1990), p. 12, asserts that "Influenced by Henry More's Christian Neoplatonism, his own extensive alchemical investigations, and his own commitment to a voluntarist notion of divine activity, Newton rejected the brute mechanisms of traditional mechanical philosophies, infusing the inert world of matter with the activity of the divine will – either directly through the hand of God or indirectly through active principles, which gave the world a structure and order that evinced providential choice rather than blind mechanical necessity."

68   Gunton, *The Triune Creator*, p. 129.

69   Gunton, *Yesterday & Today: A Study of Continuities in Christology* (Grand Rapids: Eerdmans, 1983), p. 120.

64 *Transcendence and Spatiality of the Triune Creator*

perience of the world possible. Kant's concepts of space and time are not derived from experience, as Newton appears to have held, but brought to it, that is, imposed upon it by the human mind. Following the thought of Newton, Gunton argues, Kant believed that the mind is *bound* to order its experience according to the mechanical laws of motion.[70] One of the main effects of Kant's idealism was thus also to drive God out of the world.

> Kant never denied the existence of that world beyond; what he denied was that there could be real knowledge of it. God is useful, indispensable even, as a regulative concept, giving us an idea of the unity of things and thus a motive for science. But the overall effect as the years passed was progressively to exclude God from the world.[71]

A symbolic event of this outcome is the famous story of the French scientist Laplace. When he presented his account of the solar system to Napoleon, the emperor asked why there was no place in it for God, unlike in Newton's model. Laplace replied, "I have no need of that hypothesis."[72]

According to Torrance, with its powerful ingredient of neo-Platonism, Augustinian tradition dominated the Middle Ages. In this tradition, this world or universe was regarded as "a sacramental macrocosm." The world of physical and visible creation was held to be the counterpart in time to eternal and heavenly patterns. Thus, since "the world of nature was looked at only sacramentally, i.e. looked through toward God and the eternal realities," Torrance holds, "as such the world had no significance in itself, or only significance in so far as it participated in divine and eternal patterns."[73] In this situation the development of empirical science was impossible.

However, with the Reformation, "there emerged a new outlook involving the primacy of Grace and the rethinking of its meaning as the turning of God toward the world,"[74] Torrance maintains. If in the pre-

---

70   Gunton, *The Triune Creator*, p. 132, says, "This is the very place where Kant canonized the thought of Newton."
71   Gunton, *The Triune Creator*, p. 133.
72   Cf. Gunton, *The Triune Creator*, p. 133 and Worthing, *God, Creation, and Contemporary Physics*, p. 26.
73   Torrance, *Theological Science* (Edinburgh: T&T Clark, 1996), p. 66f.
74   Torrance, *Theological Science*, p. 67.

Divine Transcendence

Reformation outlook the world was interpreted in terms of its attraction toward God, in the new outlook after Reformation the world was interpreted in terms of God's action upon the world. Hence, Torrance asserts, "the way was opened up for the development of empirical science which is inhibited so long as man looks only away from the world to God to find its meaning in its participation in divine patterns."[75] But Torrance points out, "once this outlook is established and the primacy of Grace is undermined, there arise tendencies toward Deism, which has room for God only in the ultimate beginning of creation, or toward agnosticism, which takes seriously the purely contingent nature of all that is not God but is tempted to convert this contingency into a new necessity."[76] Thus Torrance describes "the *risk* of Deism or agnosticism" as "the price that Protestantism pays for the liberation of nature."[77] In this situation, the development of empirical science became possible.

Deism can be understood as "a rationalized version of the essentially non-trinitarian view of the relation between God and the world."[78] Deists hold that the only believable God is one discovered by reason alone, and whose relation to the world can be none other than that of machine-maker to machine. After creation, there is no real relation between God and the world in deism. Gunton suggests that a more fully trinitarian account of creation would read as follows:

> we are freed at once from mechanism and pantheism, as well as the latter's near relation, panentheism, because we have a way of establishing the reality of the world as world, while yet it remains in relation to the one who made it. It follows first that the world of space and time is what it is by virtue of its relation-in-otherness with the creator. Space and time are not continuous with God, which means that they are, as created realities, in some way functions of there being a created being.[79]

This saying is consistent with Pannenberg's contention that, in order to describe the relationship of God's transcendence and immanence in creation and in the history of salvation, today's theology of creation should

---

75 Torrance, *Theological Science*, p. 67.
76 Torrance, *Theological Science*, p. 67.
77 Torrance, *Theological Science*, p. 67, note 1.
78 C. E. Gunton, *The Triune Creator*, p. 126f.
79 C. E. Gunton, *The Triune Creator*, p. 143.

66 *Transcendence and Spatiality of the Triune Creator*

use the doctrine of the Trinity.[80] In process theology, which contends for panentheism, we do not find much discussion on the doctrine of the Trinity.[81] Process theologians are more concerned about the dipolarity of God than the Trinitarian idea. Process theology is thus "a detour away from the main direction of Trinitarian discussion."[82] The same can be said of deism. Deism is also a non-trinitarian view. Yet, the doctrine of the Trinity is arguably the most useful tool to hold together both divine transcendence and immanence.

5.   Transcendence and Spatiality

Contrary to the view of classical theists such as Augustine, Anselm and Aquinas, that "God's eternal existence is to be understood as timelessness," so-called temporalist theists would assert that God's eternal existence is to be interpreted as everlastingness rather than as timelessness. Nash identifies the following temporalist theists with their views. Duns Scotus and William of Occam were the first to challenge the atemporalist view. Samuel Clark, a theological advocate of Newton, and Jonathan Edwards, one of the great American philosophers, later proposed that God is everlasting rather than timeless. Contemporary proponents of the temporalist position are as follows: Nelson Pike, Nicholas Wolterstorff, Richard Swinburne, William Rowe, and Anthony Kenny.[83]

Gunton definitely rejects this complementary tendency to read our time into God, even though it is more fashionable today. He presents the reason for his rejection as follows: "if we view the relation of God and

---

80   See Pannenberg, *Toward a Theology of Nature*, p. 65, says: "Today's Christian theology of creation will use, in distinction from Newton, the possibilities of the doctrine of the Trinity in order to describe the relationship of God's transcendence and immanence in creation and in the history of salvation."

81   Cf. John B. Cobb, Jr. & David Ray Griffin, *Process Theology: An Introductory Exposition* (Philadelphia: The Westminster Press, 1976), p. 110, state that "process theology is not interested in formulating distinctions within God for the sake of conforming with the traditional Trinitarian notions."

82   T. Peters, *GOD as Trinity: Relationality and Temporality in Divine Life* (Louisville: Westminster, 1993), 114.

83   Nash, *The Concept of God*, p. 73f.

Divine Transcendence 67

the world from the perspective of the created order, the danger is that God will be conceived by a process of abstraction from the world."[84] A number of more moderate accounts of what has been called temporalistic theism all founder, Gunton avers, on a version of Augustine's argument that, if God is the creator of time, then time belongs to the structure of created things, and accordingly not to God himself. Gunton asserts that "the doctrine of creation, in teaching that it is one thing to be God, another to be the created order, rules out a simple, or even a complex, reading up of temporal reality into God."[85] Nevertheless, Gunton highlights the positive relation of God to time in his noting the works of Christ and the Spirit. Thus Gunton concludes, "we should accept neither the timelessness nor the temporality of the being of God."[86] In this sense, Gunton does not consent totally to Augustine's view of time. Unlike Augustine, for example, Gunton does not think that eternity is timelessness. He proposes "a perichoresis, a simultaneous interaction, of the temporal and the eternal."[87] The eternal is not conceived as a time-denying but as a time-embracing reality.

God is the creator of space and time which are the conditions of creatures. In this sense, I agree with Gunton's view that we should not read our time into God. If we understand temporality as a kind of limitation, then God is not a temporal being. For the creator must not be tied to the creaturely structures of time and space. For Augustine, time is also a creature of God. God the creator, thus, is transcendent over the temporal world. In this sense, God is not a temporal being like other creatures. However, God interacts in the world of space and time which he has created. If God is an atemporal being, we have a hard time explaining his immanence, or presence, in the world. Therefore, we might presuppose a kind of temporality of God. In this sense, God cannot be just an atemporal being, even though we can acknowledge that God is not a temporal being.

This view of Gunton on God and time can be applied to the present discussion of God and space: God is the creator of space and time. Therefore, we should not read up from our spatial reality to deduce the meaning of God's spatiality. God is not a spatial being, for the creator

---

84   Gunton, *The Triune Creator*, p. 91f.
85   Gunton, *The Triune Creator*, p. 92.
86   Gunton, *The Triune Creator*, p. 92.
87   Gunton, *Yesterday and Today*, p. 130.

68 *Transcendence and Spatiality of the Triune Creator*

cannot be tied to the structures of space created by him. But the positive relation of God to space on the grounds of the work of the Son and the Spirit may be asserted. Thus we reach the conclusion that *we should accept neither the spacelessness nor the spatiality of the being of God.* Gunton proposes that we should stop our discussion at this point. Anything beyond this statement is necessarily mere speculation. Thus, he maintains, "we cannot but be apophatic, because no more is shown us."[88]

David Ford defines the term "apophatic" as follows: "Relating to what is beyond expression in speech; more specifically, relating to the method of negative theology which stresses the transcendence of God over all human language and categories, and prefers forms of reference which say what God is not."[89] Torrance, contrary to Gunton, opposes the apophatic position. Grappling with the Trinity, he says:

> This does not mean that when we reach this point, at the threshold of the Trinity *ad intra*, theological activity simply breaks off, perhaps in favour of some kind of merely apophatic or negative contemplation. While the Triune God is certainly more to be adored than expressed, the ultimate Rationality, as well as the sheer Majesty of God's self-revelation, and above all the Love of God, will not allow us to desist.[90]

Like Torrance, I am not satisfied with the apophatic conclusion of Gunton, for we have some distinct biblical and theological concepts to support a kind of spatiality of God. We are taught that God created the world of space and time as well as heavens. God promised a land which lies in the so-called Middle East to the nation of Israel. God was "in" the Temple which He ordered built in one location. Furthermore, God became incarnate in the history of space and time. Finally, Jesus was resurrected in a bodily form, and God promises us a bodily resurrection at the eschaton. And ultimately, in the present time, the resurrected Christ is, in some sense, really present in the sacraments by the Spirit. In the Gospel according to John we find the concept of the mutual indwelling between God and believers. In the Pauline epistles we find that we are "in" Christ.

---

88 Gunton, *The Triune Creator*, p. 92. In this context, however, we should pay attention to the fact that Gunton leaves an eschatological reserve that "we require an eschatological as well as a relational conception of time" (ibid.).

89 David F. Ford, ed., *The Modern Theologians*, p. 734.

90 Torrance, *The Christian Doctrine of God*, p. 111.

Divine Transcendence 69

Of course we cannot use directly these biblical statements to prove the spatiality of God. But, they do suffice to compel believers to consider moving beyond the apophatic position. We may argue that, even though God is not a being in space and time like other creatures, God is in some sense spatial, or has a kind of spatiality. At this point, William A. Dyrness's three categories of relationship, agency, and embodiment serve as useful descriptions.[91] According to him, these categories, which are grounded in the triune presence of God, provide windows by which illumine our human life in the world. We may apply his view to suggest a positive relation between space and God. Through these three categories, we may assert, we can approach the spatiality of God.

Among the eight biblical pointers mentioned above, the incarnation of the Son is crucial to this discussion. According to Torrance, "incarnation means that He by whom all things are comprehended and contained by assuming a body made room for Himself in our physical existence, yet without being contained, confined or circumscribed in place as in a vessel. He was wholly present in the body and yet wholly present everywhere, for He became man without ceasing to be God."[92] Such a statement does not make sense on the basis of the Greek concept of space as a receptacle or container. This statement depends for its coherence upon a notion of space according to which the relation between God and space is not itself a spatial relation.[93] Thus Andrew Purves, a former student of Torrance, contends, that for Torrance "God, the creator of space, stands in a non-spatial relation to creation, yet God has entered into space in such a way that all his relations with us occur within spatial (and temporal) reality."[94] Torrance argues, "while the Incarnation does not mean that God is limited by space and time, it asserts the reality of space and time for God in the actuality of His relations with us, and at the same time binds us to space and time in all our relations with Him."[95] Together with

---

91   Dyrness, *The Earth is God's*, p. xii.
92   Torrance, *Space, Time and Incarnation*, p. 13.
93   Torrance, *Space, Time and Incarnation*, p. 2. "God is the transcendent Creator of the whole realm of space and stands in a creative, not a spatial or a temporal relation, to it" (ibid., p. 3).
94   Andrew Purves, "The Christology of Thomas F. Torrance," in *The Promise of Trinitarian Theology: Theologians in Dialogue with Thomas F. Torrance*, ed. Elmer Colyer (Lanham, MD: Rowman & Littlefield, 2001), p. 56.
95   Torrance, *Space, Time and Incarnation*, p. 67.

70   *Transcendence and Spatiality of the Triune Creator*

the creation, he states, the incarnation forms "the great axis in God's relation with the world of space and time." Apart from creation and incarnation, our understanding of God and the world can only lose meaning.[96]

As stated in the previous section, on the one hand, the doctrine of creation of the world out of nothing indicates the absolute transcendence of God. On the other hand, however, it has troubled dualists like Marcion, who would hold to a radical antithesis between the creator God and the redeemer God.[97] Furthermore, in Christian theology the contingency or materiality of the created world need not be seen as a defect of being. Contrarily, citing Torrance, Gunton indicates, "'contingent rationality' is a quest for a rationality inhering in the order of space and time, not beyond it. This, it is claimed, is the unique gift of the Christian doctrine of creation. The material world is contingent but rational."[98] Dyrness similarly points out that the doctrine of creation was one of the early church's first contributions to theology, which was born in the struggle against the Gnostics. Citing Calvin, Dyrness writes, "Calvin [...] argued that creation is the theater for the display of God's glory, going so far in one place as to say, 'I confess, of course, that it can be said reverently, provided that it proceeds from a reverent mind, that nature is God.'"[99]

---

96   Torrance, *Space, Time and Incarnation*, p. 68.
97   Cf. Alister E. McGrath, *T. F. Torrance: An Intellectual Biography* (Edinburgh: T&T Clark, 1999), p. 158, says: "Marcion's radical antithesis between the creator God and the redeemer God illustrates this dualism, as does Arius' conception of the manner in which the divine and the human 'relate to one another merely tangentially without in any way intersecting one another'. For Torrance, Einstein's unitary way of approaching nature was characterized by its integration of empirical and theoretical factors, so that ontology and epistemology are 'wedded together.'"
98   Gunton, *The Triune Creator*, p. 112f. My argument here follows McGrath, *T. F. Torrance: An Intellectual Biography*, p. 221, footnote 77. Moltmann also acknowledges Torrance's contribution to this area: "We have to thank T. Torrance for his working out of the concept of a 'contingent order' of contingent events. Cf. *Christian Theology and Scientific Culture*, pp. 16ff. He has thereby independently developed M. Polanyi's approaches in *Knowing and Being*, London and Chicago 1969" (Moltmann, *God in Creation*, p. 347, n. 27).
99   Dyrness, *The Earth is God's*, p. 27. This saying is quoted from Calvin, *Institutes*, 1.5.5. In an endnote, Dyrness adds the following: "Calvin goes on to qualify the statement, showing that it is the sovereignty of God that he wishes to underline" (ibid., p. 171, note 6).

Divine Transcendence

The land of Canaan is not the boundary of God's power. Yet, it can be regarded as part of the covenant promise. Therefore, the promise of the land to the Israelites was never to imply that Yahweh is simply a local deity. By contrast the pagan gods of antiquity were usually limited in space, in function, or in both. Thus, Gerald Bray notes, "the God of the Bible is thus distinguished from all pagan deities both by his omnipotence and by his omnipresence. It is true that he was the God of a particular people, but even in the Old Testament his lordship was not confined either to them or to the land of Israel, as Jonah discovered to his cost."[100] The God of the Bible is infinite. This statement means He is free from all creaturely limitations, surely. But it does not imply that God is not present in this creaturely world. He is everywhere. Citing Barth's assertion that God is infinite and finite, Bloesch contends as follows: "We may call God 'infinite, measureless, limitless, spaceless and timeless,' but 'this does not mean that we will try to exclude, deny or even question that He is the One who in His whole action posits beginning and end, measure and limit, space and time.' God is infinite but he can enter into the finite, he can take the finite into himself."[101] Concerning the meaning of omnipresence, Bloesch says, it does not signify spacelessness but God's freedom to be in space: "He is present to all his creatures, but there is no identity between God and the creature."[102]

In the Old Testament period, God was present in the Temple which He ordered to be built on Mount Zion. Solomon prayed at the dedication ceremony of the Temple: "But will God really dwell on earth? The heavens, even the highest heaven, cannot contain you. How much less this temple I have built!" (1 Ki. 8:27). Regarding this issue, Horst D. Preuss asserts the following: "there is the statement that YHWH dwells in heaven (Pss. 2:4; 123:1; Deut. 4:36; and 26:15), and from there he descends to the Sinai or to another place. From there he looks below (Pss. 33:13f. and 102:20); he also descends from his heavenly dwelling to the temple (cf. The shout in Hab. 3:20; Zeph. 1:7; and Zech. 2:17), where he then takes up his temporary presence [...] The 'throne of YHWH that towers up to the heaven' is established in the Jerusalem

---

100   Gerald Bray, *The Doctrine of God* (Downers Grove, IL: IVP, 1993), p. 86.
101   Donald G. Bloesch, *God the Almighty* (Downers Grove, IL: IVP, 1995), p. 85.
102   Bloesch, *God the Almighty*, p. 87.

72 *Transcendence and Spatiality of the Triune Creator*

temple."[103] This Jerusalem temple of the Old Testament anticipates the incarnation of Jesus in the New Testament. The Gospel according to John says, "The Word became flesh and made his dwelling among us. We have seen his glory, the glory of the One and Only, who came from the Father, full of grace and truth" (Jn. 1:14). The Greek word for "made his dwelling" is ἐσκήνωσεν, which means "to live in a tent, to settle" in order to take up one's temporary dwelling place.[104]

In the Gospel according to John we find the concept of mutual indwelling between God the Father and God the Son *and* God and believers. Donald Guthrie explains this theme:

> In the bread discourse the person who eats Christ's flesh and drinks his blood is said to abide in Christ ('abide in me, and I in him', Jn. 6:56). The idea of abiding is especially frequent in the farewell discourse. In John 14:10 Jesus asks, 'Do you not believe that I am in the Father and the Father in me?' He promises that his disciples would know that he was in the Father and the Father in him (Jn. 14:20). He prays for his disciples 'that they may all be one; even as thou, Father, art in me, and I in thee, that they also may be in us, so that the world may believe that thou hast sent me' (Jn. 17:21). In these passages the union between the Father and the Son is seen to be the pattern for the believer's life in God. In the vine allegory in John 15 the idea of abiding is expressed in the double form 'Abide in me and I in you' (Jn. 15:4; cf. 15:5).[105]

It is from these passages that Christian theologians have developed the doctrine of perichoresis. Elmer M. Colyer comments on perichoresis as follows: "as developed by other Nicene theologians, *perichoresis* refers to the reciprocal relations between the Father, the Son and the Holy Spirit in which they mutually indwell, coinhere, inexist and wholly contain one another without in any way diminishing the persons and their real distinctions."[106] Following in this line of thought, Moltmann has boldly set forth his concept of perichoretic space. This mutual indwelling

---

103 Horst D. Preuss, *Old Testament Theology*, vol. 1, trans. Leo G. Perdue (Louisville, KY: Westminster John Knox Press, 1995), p. 251.
104 Fritz Rienecker, *A Linguistic Key to the Greek New Testament* (Grand Rapids: Zondervan, 1980), p. 218.
105 Donald Guthrie, *New Testament Theology* (Downers Grove: InterVarsity Press, 1981), p. 641f.
106 E. M. Colyer, *How to Read T. F. Torrance*, p. 314.

Divine Transcendence 73

of John, Guthrie holds, provides "valuable parallels with the 'in Christ' idea in Paul's epistles and will furnish some light on its meaning."[107]

In the Pauline epistles we read that we are "in Christ." Sometimes Paul speaks of the believer dwelling in Christ or in the Spirit, and sometimes of Christ or the Spirit indwelling the believer. According to Guthrie, "These are complementary, not contradictory, concepts."[108] Usually the phrasing is used within the context of describing new life in Christ.[109] After indicating that "the expression 'in Christ' is one of Paul's most characteristic formulations and its precise meaning has been vigorously debated," George Eldon Ladd introduces Deissmann's view on the phrase. Deissmann emphasizes its "mystical" dimension. Regarding this interpretation, Ladd comments,

> Basic to Deismann's interpretation is the identification of Christ and the Spirit (2 Cor. 3:17). The "Spirit-Christ" has a body that is not earthly or material, but consists of the divine effulgence. The Spirit-Christ is the Christian's new environment. It is analogous to the air. As we are in the air and the air is in us, so we are in Christ and Christ is in us [...] This very idea will seem intolerable to people unfamiliar with ancient ways of thought who conceive of "spiritual" as *ipso facto* nonmaterial. However, the ancient world had different thought categories. "Spirit" could be understood in terms of fine invisible matter that could interpenetrate all visible forms of matter.[110]

This term "in Christ" is used to remark on conscious communion with Christ. It also has a corporate dimension. Thus Ladd holds that "believers are in Christ not only as individuals but as a people."[111]

Jesus was resurrected in a bodily form, and so also God promises followers of Christ a bodily resurrection at the eschaton. For Torrance, "every kind of deistic dualism between God and our world is rejected by the resurrection of Christ."[112] Quoting Augustine's view, he writes: "even after the resurrection the body of Christ was called 'flesh' (Luke 24: 39)."[113] Torrance contends that "Since man is the concrete reality he is,

---

107 Guthrie, *New Testament Theology*, p. 641.
108 Guthrie, *New Testament Theology*, p. 647.
109 Cf. Guthrie, *New Testament Theology*, p. 641 and G. E. Ladd, *A Theology of the New Testament* (Grand Rapids: Eerdmans, 1994, Revised Edition), p. 521.
110 Ladd, *A Theology of the New Testament*, p. 523.
111 Ladd, *A Theology of the New Testament*, p. 524.
112 Torrance, *Space, Time and Resurrection*, p. 65.
113 Torrance, *Space, Time and Resurrection*, p. 75, footnote 13.

74   *Transcendence and Spatiality of the Triune Creator*

resurrection of man in the nature of the case can be only bodily resurrection – any 'resurrection' that is not bodily is surely a contradiction in terms."[114] The central message of Christianity about the future of human beings is the bodily resurrection. This thought should suggest the parameters and crucial elements of any discussion of the spatiality of God.

The resurrected Christ is, in some sense, really present in the Sacraments by the Spirit. Regarding the entry for "sacramentum," in the *Dictionary of Latin and Greek Theological Terms*, Richard Muller defines the term as follows: "in churchly usage, a holy rite that is both a sign and a means of grace [...] The sacraments are also defined by the scholastics as the visible Word of God, distinct but not separate from the audible Word or Holy Spirit. In the traditional Augustinian definition, a sacrament is a visible sign of an invisible grace."[115] In the Eucharist, Christ is present. As stated previously in this study, the debate over the mode of Christ's presence in the Eucharist may be formulated as in part a contest between competing views of space and time.

6. Conclusion

To refute the threats of pagan pantheism, the traditional Christian doctrine of God has emphasized divine transcendence. The distinction between the creator and creatures has been seen as more important than anything. Thus, even though classical theism does not abandon divine immanence or presence in this world of space and time, it has not given

---

114 Torrance, *Space, Time and Resurrection*, p. 83. In the introduction to *Space, Time and Resurrection*, Torrance describes an encounter with Barth during the last visit to him: "Then I ventured to say that unless that starting point was closely bound up with *the incarnation*, it might be only too easy, judging from many of our contemporaries and even some of his former students, to think of the resurrection after all in a rather docetic way, lacking concrete ontological reality. But at that remark, Barth leaned over to me and said with considerable force, which I shall never forget, *'Wohlverstanden, leibliche Auferstehung'* – 'Mark well, bodily resurrection'" (ibid., p. xi).

115 Richard A. Muller, *Dictionary of Latin and Greek Theological Terms: Drawn Principally from Protestant Scholastic Theology* (Grand Rapids: Baker Books, 1985), p. 267.

Divine Transcendence

75

due significance to it. In comparison to this leaning to the divine transcendence of traditional theism, our times may be described as an age of immanence. The dominant concern of modern Christian theology has been with divine immanence. Yet, the relation of divine transcendence to divine immanence need not be an either-or relation. In this age of the crisis about divine transcendence, we should avoid a solution of "immanent transcendence," while noting that "absolute transcendence" or "out-and-out" kind of transcendence does not fit the biblical data either. If absolute transcendence is the only option, there can be no revelation at all, and we cannot know God at all.

In opposition to theism and pantheism, panentheists have been suggested to argue for a balance between the two. However, panentheism does not fully confirm divine transcendence. Panentheists usually identify deism with classical theism, since deism and classical theism share an emphasis on divine transcendence. But, unlike deism, classical theism admits of divine immanence. Otherwise, classical theism could not affirm the incarnation of the Son. Classical theism indeed contends for the interaction of the creator God with the world.

In orthodox Christian theologies, God is not only transcendent over the world but also immanent within it. God is present and interacts in this world. God has a positive relationship with the world of space and time. Thus, I would maintain that divine transcendence should not necessarily exclude divine spatiality nor temporality. But we should caution that we do not read our time and space into God. God is the creator of time and space. In this sense, like Gunton, I would agree with Augustine's view that God is not a temporal being like any other creatures. At the same time, however, I do not agree with Augustine's view that God is timeless or spaceless. God has a kind of temporality and spatiality. In closing this chapter on the divine transcendence, I am satisfied with the passive, or negative, conclusion that transcendence does not necessarily exclude divine spatiality. In the next chapter, I will examine theological thoughts about God and space of three contemporary theologians, Thomas F. Torrance, Wolfhart Pannenberg, and Jürgen Moltmann, in order to suggest a more positive statement on the spatiality of God.

Chapter IV

God, Space, and Spatiality-Case Studies

1. Introduction

Chapter III discussed some possibilities of accounting for God's spatiality without denying his transcendence. Noting that the spirit of our times is preoccupied with divine immanence in the world, I argued that theologians today must affirm God's transcendence beyond the world of space and time. With the emergence of the Newtonian world-view, deism presented a threat to Christian theism. As a reaction to the God of deism, who only transcends but cannot interact in this universe, panentheism has recently offered a popular option for many Christian theologians, including process theologians, feminist theologians, and some liberation theologians. Panentheists assert that, whereas classical theism overemphasizes divine transcendence, pantheism focuses on divine immanence. Panentheists aver that their position can hold together both God's transcendence and immanence effectively as opposed to classical theism or pantheism. But the transcendence of panentheism is, in Anderson's term, an "immanent transcendence,"[1] and thus falls short of offering a valid alternative to the imbalances of classical theism.

Panentheists usually commit the mistake of identifying classical theism with deism. They criticize classical theism for unduly emphasizing divine transcendence. But in classical theism both divine transcendence and immanence are emphasized. Classical theism is not to be identified with deism. Nevertheless, the panentheists' charge against classical theism contains an element of truth. In an overreaction to the danger of pan-

---

1   R. S. Anderson, *Historical Transcendence and the Reality of God: A Christological Critique* (Grand Rapids: Eerdmans, 1975), p. 24, presents the following explanation: "The 'immanent' kind of transcendence [...] conceives no fixed point external to the contingent world, but transcending relationships within the process of infinite possibility."

78   *Transcendence and Spatiality of the Triune Creator*

theism, classical theism has indeed failed to account for divine immanence. Therefore, along with overcoming panentheism, the most pressing
issue for contemporary Christian theologians is "how to avoid de facto
deism – not merely by calling it unorthodox and expressing their dislike
of it."[2] The transcendence of deism is, again in Anderson's term, an "out-
and-out transcendence."[3]

By analogy, with Gunton's apophatic position on divine temporality,
we may begin to suggest a way to speak of God's spatiality: we should
accept neither the spacelessness nor the spatiality of the being of God.
Yet, I would also seek to go beyond this apophatic position to suggest
that there is some room for speaking of God's spatiality in a way which
is not contradictory to biblical and theological data on divine transcendence. Through the argument of the previous chapter, therefore, I
reached the conclusion that the divine spatiality does not necessarily
contradict the transcendence of God.

In this chapter I will do a kind of case study about their theological
thoughts on divine spatiality of three theologians. All of them are involved in the dialogue between theology and science. Addressing the
dialogue between science and religion, John Polkinghorne indicates that
the most grievous absence from this conversation is that of theologians.
For fruitful development in the future, he expresses earnest hope for the
participation of theologians in this dialogue. "Because God is the ground
of all that is," Polkinghorne says, "all human knowledge is the concern
of the theologian[s]." However, he acknowledges the difficulties they
have. Polkinghorne points out two difficulties of theologians especially:
First, "books are many and life is short, and much science is formidably
technical in appearance, although many of its concepts are capable of at
least partial expression in lay language." Furthermore, "much twentieth-
century theology has been either fideistic (Barth) or existential

---

2   Philip Clayton, *God and Contemporary Science* (Grand Rapids: Eerdmans, 1997),
    p. 192.
3   Anderson, *Historical Transcendence and the Reality of God*, p. 24, states: "The
    'out-and-out' kind of transcendence is plain talk for saying that there is an ultimate
    reality which has its own existence independent of and beyond a contingent and fi-
    nite world. That is, it is a concept of a fixed point outside of the process by which
    all moving points take their reckoning."

God, Space, and Spatiality-Case Studies

(Bultmann) in tone, conducted from within ghettoes walled off from scientific culture."[4]

As some honorable exceptions to this policy of keeping theology at a distance from science, Polkinghorne presents three theologians: Torrance, Pannenberg, and Moltmann. According to Polkinghorne, one of the greatest contributions of Torrance in this dialogue between science and religion is "the insistence that components of reality, from electromagnetic fields to God, are known in ways that accord with their natures and that we cannot determine beforehand what these epistemological modes will be."[5] Yet, at the same time, Polkinghorne points out Torrance's limitations. James Clerk Maxwell and Albert Einstein, scientific heroes of Torrance, he asserts, are the last of the ancients rather than the first of the moderns. As a result, Polkinghorne contends, "we do not find in his thought, for example, much engagement with the veiled elusiveness of the quantum world."[6]

After indicating that Pannenberg's major writing in this area has been concerned with the philosophy of science, rather than with science itself, Polkinghorne introduces two particular themes of Pannenberg in the content of science. The first one is Pannenberg's preoccupation with the concept of inertia which, according to Polkinghorne, is a rather old-fashioned idea. He states, "Pannenberg seems to treat as if it were a kind of burdensome constraint on the freedom of creation, or even of the Creator."[7] The second one is Pannenberg's idea of a field. Pannenberg supposes this concept of field, in its spatiality spread out character, to offer a metaphor, or even perhaps more than a metaphor, for omnipresent Spirit. Concerning this idea, however, Polkinghorne criticizes, "To a physicist this seems rather quaint."[8]

---

4   J. Polkinghorne, *Belief in God in an Age of Science* (New Haven & London: Yale University Press, 1998), p. 80.

5   Polkinghorne, *Belief in God in an Age of Science*, p. 81. Elsewhere he writes, "Thomas Torrance has been particularly emphatic in stressing the necessarily open dependence of knowledge on the nature of that which is known" [See Polkinghorne, *The Faith of a Physicist* (Minneapolis: Fortress Press, 1996), p. 33].

6   Polkinghorne, *Belief in God in an Age of Science*, p. 81.

7   Polkinghorne, *Belief in God in an Age of Science*, p. 82.

8   Polkinghorne, *Belief in God in an Age of Science*, p. 82.

80   *Transcendence and Spatiality of the Triune Creator*

In several places, Polkinghorne expresses sympathy with Moltmann's thought.[9] But in the area of dialogue between science and religion, he comments on Moltmann's weaknesses. Citing Moltmann, Polkinghorne states, "In relation to a discussion of the 'Space of Creation,' he acknowledges only a future willingness: 'I hope to be able to develop this subject further elsewhere in the light of new scientific conception about space-time continuum.'"[10] However, according to Polkinghorne, so far Moltmann's execution has fallen short of this promise. Polkinghorne contends that, "It is certainly not possible to talk seriously today about space without taking relativity into account."[11]

In each section of this chapter, I will examine the views on God, space, and God's spatiality of these three contemporary theologians. The purpose of the present chapter is to further examine the thoughts of these theologians on divine spatiality in the light of the divine transcendence proposed in the previous chapter. Neither "out-and-out transcendence" nor "immanent transcendence" can be appropriate to Christian theology. In other words, neither deism nor panentheism is a suitable alternative to Christian theism. Torrance develops a notion of divine temporality. But he does not contend for divine spatiality. Because of his concern for divine transcendence, Torrance appears to be hesitant to argue for God's spatiality. Pannenberg explains the activity of the Spirit in terms of the field concept of modern physics. This attempt has value in expressing the continuing creative presence of God. Using this concept of field not only as a metaphor, however, Pannenberg goes too far toward a panentheistic idea. Finally, Moltmann has presented a view of the doctrine of creation and the Shekinah. His doctrine of creation as divine *zimzum* implies panentheism. He develops various dimensions of Christian theology. But his panentheism inhibits him from adequately describing di-

---

9   Concerning Moltmann's explanation of *nihil*, see Polkinghorne, *Science and Providence: God's Interaction with the World* (London: SPCK, 1989), p. 22 and *Science and Creation: The Search for Understanding* (London: SPCK, 1988), pp. 61–62. Concerning "heaven" and "earth," see *Science and Creation: The search for Understanding*, pp. 79–80 and *Reason and Reality: The Relationship between Science and Theology* (Philadelphia: Trinity Press International, 1991), p. 42. Concerning Moltmann's social trinitarianism, see *Science and Providence*, p. 88. And finally concerning the problem of suffering, see *Science and Theology: An Introduction* (London/ Minneapolis: SPCK/ Fortress Press, 1998), p. 112.

10   Polkinghorne, *Belief in God in an Age of Science*, p. 81.

11   Polkinghorne, *Belief in God in an Age of Science*, p. 81.

God, Space, and Spatiality-Case Studies

vine transcendence. He sacrifices God's transcendence on behalf of divine immanence or spatiality.

2. Torrance-Redemption of Space

This section deals with Torrance's thought on God and space. Subsection 1 examines his view on the relation between God and the world, in which he seeks to overcome dualism. Torrance insists on a dynamic interaction between God and the world. Subsection 2 treats Torrance's contention on the redemption of space and time. He distinguishes various concepts of space and time and seems to differentiate God's space-time from our space-time. Subsection 3 examines Torrance's view on God's temporality. He definitely defends a notion of divine temporality. He acknowledges the temporal order in the being of God. Subsection 4 critically evaluates these thoughts of Torrance. Torrance's views offer much grist for a discussion on God's spatiality. However, he has not developed his suggestions fully.

2.1 God and the World

Torrance rejects ancient and modern forms of dualism. This rejection of dualism is crucial to his theological perspective on the God-world relation and his critical realist epistemology. Concerning this point, Elmer M. Colyer says, "Torrance does not begin by rejecting dualism, and then develop his God-world relation and epistemology. Rather his God-world relation and epistemology arise out of his theological investigation and lead to his rejection of dualism."[12]

Torrance mainly focuses his refutation on cosmological and epistemological dualism. In cosmological dualism, "God is not thought of as interacting with the universe of space and time but as inertially related to

---

12   Elmer M. Colyer, *How to Read T. F. Torrance: Understanding His Trinitarian and Scientific Theology* (Downers Grove: InterVarsity Press, 2001), p. 57, note 9.

82   *Transcendence and Spatiality of the Triune Creator*

it or as deistically detached from it."[13] This posits a separation between God and the world. Torrance sees this separation and repudiates it in both Greco-Roman philosophy in the cultural milieu of the early church and Newtonian, deistic and other dualist perspectives of the modern period. In this dualistic framework, there is a deep chasm between God and the world.

In epistemological dualism, there is the presupposition of a disjunction between the human knower and the reality that the human subject seeks to know. The representative of this position is Kantian dualism, in which we cannot really know reality or the thing in itself *(Ding an sich)*. Thus, Torrance says,

> [T]here is no factual knowledge of God except where He has condescended to reveal Himself in His objectivity. We cannot know Him by transcending or going behind His objectivity, for that would be to go where no God is to be found and where there is no divine object for our knowledge. We know God, in the proper rational sense of knowledge, only in His objectivity, not by seeking non-objective knowledge of Him.[14]

Concerning the relationship between theological and natural science, Torrance asserts that they operate within the same medium of space and time. According to him, space and time are the bearers of contingent order or intelligibility in which all created realities share. Torrance holds that "while the two sciences inevitably overlap within the space-time of this world, they move in different directions, one to investigate creaturely relations out of themselves, apart from God, and the other to inquire of God who transcends all creaturely relations and makes himself known through his Word as the Lord of all space and time."[15]

Dualism has been detrimental to both theology and natural science, Torrance believes. In theology, one evil influence of dualism is the deistic disjunction between God and the universe. As a result, God is not

---

13   Thomas F. Torrance, *Reality and Evangelical Theology: The Realism of Christian Revelation* (Downers Grove: InterVarsity Press, 1999), p. 32.

14   Torrance, *Theological Science*, p. 55. Colyer states, "within epistemic dualism it is difficult or impossible to know much of anything about Jesus in himself, for what we have in the New Testament is Jesus as he has been appreciated and interpreted by various authors/communities in light of their perspectives and agendas" (E. M. Colyer, *How to Read T. F. Torrance*, p. 58, note 10).

15   Torrance, *Reality and Evangelical Theology*, p. 30.

God, Space, and Spatiality-Case Studies

conceived as acting objectively within space and time, and theological and natural science are kept entirely apart. In natural science, Torrance says, "a dualism forces upon it a mechanistic conception of the universe as a closed gravitational system of bodies in motion which are externally connected through cause and effect, in which the idea of God has no place at all, with the result that theology tends to be secularized out of the scientific and academic realm as an anachronistic irrelevance."[16]

Torrance identifies four aspects of the new scientific outlook which have been fruitful for dialogue between science and theology. Torrance refers to a basic change in the "concept of reality" as the first point. He explains as follows:

> This has to do with the transition from the earlier concept of reality, which since the days of Galileo and Newton was identified with what is causally necessary and quantifiable, the world of 'real, mathematical time and space,' as Newton called it, in contrast to 'the apparent and relative time and space' of our ordinary experience, to a new concept of reality in which that kind of dichotomy is transcended and in which structure and matter, or the theoretical and empirical components of knowledge, are inseparably one.[17]

In comparison to the older view of reality, in which its analyzed particulars (atoms, particles, etc.) were conceived of as being externally and invariably connected in terms of causes, "this is a dynamic view of the world as a continuous integrated manifold of fields of force in which relations between bodies are just as ontologically real as the bodies themselves, for it is in their interrelations and transformations that things are found to be what and as and when they are."[18]

What Torrance identifies as the second of the new scientific outlook is its "relational concept of space and time." This concept of space and time is not actually for Christianity, for it was assumed by the Greek patristic theologians. By this relational conception they sought to think out the interrelation between the incarnation and the creation, or the activity of God within space and time and his creative transcendence over all space and time. Hence, Torrance asserts, "it is not surprising that this

---

16   Torrance, *Space, Time and Resurrection*, p. 180f.
17   Torrance, *Space, Time and Resurrection*, p. 184.
18   Torrance, *Space, Time and Resurrection*, p. 185.

84   *Transcendence and Spatiality of the Triune Creator*

kind of thinking finds itself very much at home today in the post-Einstein era of scientific and cosmological thought."[19]

The third aspect of modern scientific thought which has a helpful bearing upon the interrelation between theological and other scientific inquiry is the multi-levelled structure of human knowledge, upon which Einstein and Polanyi particularly have laid stress.[20] This stratified structure of our scientific knowledge of the universe usually comprises three levels of thought coordinated with one another: "the primary or basic level, which is the level of our ordinary day-to-day experience and knowledge; the secondary or scientific level; and the tertiary or metascientific level."[21] Concerning the relation among knowledge of these three levels, Torrance writes: "the concepts at the highest level are coordinated with those at the empirical level through cross-level relations; but without being grounded in that way in our actual experience and knowledge they are ultimately no more than empty, meaningless forms of thought."[22]

The last aspect of the new scientific outlook upon the universe is a stratified structure of the universe itself. "For we have to do not only with levels of knowledge but with different levels of existence or reality."[23] Alister E. McGrath evaluates this aspect of Torrance's thought: "It will be clear that this general approach is of importance in the rehabilitation of the 'spiritual' within the 'scientific.' Yet Torrance also suggests that the recognition of the 'multi-levelled structure of human knowledge' has particular relevance to individual aspects of Christian theology."[24] Torrance's most creative application of this point, McGrath thinks, relates to the resurrection:

> Traditional discussions of the resurrection had proceeded along rather Humean lines, involving debate over whether God could violate the laws of natural through miraculous acts. Torrance argues that a multi-levelled approach to the question allows the distinctive nature of the resurrection to be appreciated, while avoiding re-

---

19   Torrance, *Space, Time and Resurrection*, p. 186.
20   Torrance, *Space, Time and Resurrection*, p. 188.
21   Torrance, *Reality and Evangelical Theology*, p. 35.
22   Torrance, *Reality and Evangelical Theology*, p. 35f.
23   Torrance, *Space, Time and Resurrection*, p. 191.
24   Alister E. McGrath, *T. F. Torrance: An Intellectual Biography* (Edinburgh: T&T Clark, 1997), p. 233.

God, Space, and Spatiality-Case Studies

ducing it to a 'spiritual' event (for example, purely in the experience of the disciples) or explaining it away on purely rational grounds.[25]

Torrance indicates three ways in which two hemispheres may be related to one another: "(1) as adjacent to one another but with a clear gap between them; (2) as touching one another tangentially; and (3) as intersecting one another or overlapping with one another. (1) and (2) presuppose a dualist framework of thought, whereas (3) rejects dualism in favor of interactionism."[26] Marcion's radical antithesis between the creator God and the redeemer God illustrates an example of (1). Arius' conception of the manner in which the divine and the human "relate to one another merely tangentially without in any way intersecting one another" examples (2). And Einstein's unitary way of approaching nature instances (3).[27]

The Christian doctrine of the incarnation illustrates interactionism between God and the world. This doctrine of the incarnation is concerned with the coming of God himself into space and time and becoming one of us in space and time, without of course ceasing to be God. God interacts with the world of space and time. The incarnation is an example of God's interaction with the world. In the incarnation, Torrance asserts, God "does not abrogate space and time but on the contrary supports, qualifies, and reinforces space and time, and indeed is concerned with healing what happens within our space-time world and delivering it from disorder."[28] In the light of this, what might be the relation between God and space? The next subsection examines this question.

2.2 Redemption of Space and Time

Torrance opposes both cosmological and epistemological dualism. For cosmological dualism, that is to say, the fatal deistic disjunction between God and the world, does not allow for any real Word of God to cross the gulf between God and the creature. Epistemological dualism does not

---

25 McGrath, *T. F. Torrance*, p. 233.
26 Torrance, *Preaching Christ Today: The Gospel and Scientific Thinking* (Grand Rapids: Eerdmans, 1994), p. 51.
27 Cf. McGrath, *T. F. Torrance*, p. 158.
28 Torrance, *Preaching Christ Today*, p. 49.

86 *Transcendence and Spatiality of the Triune Creator*

permit human beings in space and time any real knowledge of God as he is in himself. God acts within the world of space and time. Thus Torrance says: "I find it absurd to think that God does not freely act within the framework of space and time, or the intelligible structures of what he has created, in making himself known to mankind."[29]

In the resurrection, the Christian faith differentiated itself so sharply from "all Hellenic or Oriental dualism which had a horror of bringing God into such a relation with concrete physical reality as the flesh or body of the incarnate and resurrected Son of God."[30] Torrance insists that "Every kind of deistic dualism between God and our world is rejected by the resurrection of Christ."[31]

However, while God is pleased to be present and to act in space and time without in any way abrogating them, Torrance maintains that God is transcendent over the world of space and time which he made. God the creator of the universe is transcendent over all time and space. Torrance regards resurrection as the redemption of space and time. He puts this as follows:

> So far we have been thinking of the resurrection as an event in space and time and yet as one that breaks redemptively through the framework of space and time, as we know it in a fallen world, but it is necessary to see that the resurrection means the redemption of space and time, for space and time are not abrogated or transcended. Rather are they healed and restored, just as our being is healed and restored through the resurrection.[32]

Our being cannot be separated from space and time, for "space and time are conditions and functions of created existence and the bearers of its order." Thus he writes, "The healing and restoring of our being carries with it the healing, restoring, reorganizing and transforming of the space and time in which we now live our lives in relation to one another and to God."[33]

For Torrance, "the kind of time we have in our fallen existence is refracted time, time that has broken loose from God, as it were, and yet not time that has been allowed to slip into sheer chaos and nothingness but

29 Torrance, *Space, Time and Resurrection*, p. 1.
30 Torrance, *Space, Time and Resurrection*, p. 81.
31 Torrance, *Space, Time and Resurrection*, p. 65.
32 Torrance, *Space, Time and Resurrection*, p. 93.
33 Torrance, *Space, Time and Resurrection*, p. 93.

God, Space, and Spatiality-Case Studies

time that is contained, upheld and overruled by God, within which he works out his redeeming purpose."[34] Far from violating or abrogating this refracted time, according to him, God redeems it. "Just as in justification the law was not destroyed but established," Torrance says, "so in the resurrection time is not annihilated but recreated, for it is taken up in Christ, sanctified in his human life and transformed in his resurrection as man."[35] Thus he explains the saying of the New Testament to "redeem the time" (Eph. 5.16): "not to live as those who are dead and asleep, rigid with the fixities of dead time, but to keep vigil as those who are already risen with Christ and wait his coming for their final release, not like Lazarus freed merely from the shrouds of the grave, but freed from the shackles of the past and all that holds us back from entering into fullness of our inheritance in the new creation."[36]

Jesus Christ really and fully became man, as we human beings are in space and time, and yet remained God the creator who transcends all creaturely being in space and time. Without "a *relational view of space and time* differentially or variationally related to God and to man," Torrance indicates, "we cannot really think the incarnation itself without falsifying it."[37] In his resurrection, "Jesus had healed and redeemed our creaturely existence from all corruption and privation of being, and every threat of death or nothingness, so that in him space and time were recreated or renewed."[38] Also in the risen Jesus, "creaturely space and time, far from being dissolved[,] are confirmed in their reality before God." In this context, Torrance refers to the dual aspects of Jesus' ascension:

> On the one hand, then, the ascension must be thought out in relation to the actual relations of space and time. On the other hand, however, the ascension must be thought of as an ascension beyond all our notions of space and time (cf. 'higher than the heavens', Heb. 7:26), and therefore as something that cannot ultimately be

---

34 Torrance, *Space, Time and Resurrection*, p. 97.
35 Torrance, *Space, Time and Resurrection*, p. 98. Torrance cites the following passage from Karl Barth's *Church Dogmatics* in his footnote: "'It is as man of his time, and not otherwise that he is Lord of time. We should lose Jesus as Lord of all time if we ignored him as a man in his own time. It is in this history – the history which is inseparable from his temporality – that the man Jesus lives and is the eternal salvation of all men in their different times.' *Church Dogmatics*, III/2, p. 440f" (ibid.).
36 Torrance, *Space, Time and Resurrection*, p. 99.
37 Torrance, *Space, Time and Resurrection*, p. 126.
38 Torrance, *Space, Time and Resurrection*, p. 127.

88 *Transcendence and Spatiality of the Triune Creator*

expressed in categories of space and time, or at least cannot be enclosed within categories of this kind.[39]

Throughout his argument, Torrance insists on God's transcendence over space and time:

God utterly transcends the boundaries of space and time, and therefore because he is beyond them he is also everywhere, for the limits of space and time which God transcends are all around us. Hence from this aspect the absence or presence of God cannot be spoken of in categories of space and time, but only when categories of space and time break off and point beyond themselves altogether to what is ineffable and inconceivable in modes of our space and time. Calvin was also right when he said that the Biblical writers never thought of the presence of God or of the ascension simply in terms of our space and time or in terms of earth and heaven.[40]

As the incarnation means that the Son of God became human without ceasing to be the transcendent God, for Torrance, so the ascension means that Jesus ascended above all space and time without ceasing to be a human being or without any diminishment of his physical or historical existence. That is the meaning of the saying that Christ is "in heaven." Thus he says, "we can speak of Christ as having a 'heavenly place' in God far beyond anything we can understand and far beyond our reach."[41]

For Torrance, the incarnation of the Son of God means the meeting of human and divine in humanity's place. The ascension of the risen Jesus means the meeting of human and divine in God's place. However, Torrance says, "through the Spirit these are not separated from one another (they were not spatially related in any case), and man's place on earth and in the space-time of this world is not abrogated, even though he meets with God in God's place."[42] Again Torrance states, "the ascension means that we cannot know God by transcending space and time, by leaping beyond the limits of our place on earth, but only by encountering God and his saving work within space and time, within our actual physical existence."[43] Thus Torrance holds, "statements regarding that ascension are *closed at man's end* (because bounded within the space-time

39   Torrance, *Space, Time and Resurrection*, p. 127f.
40   Torrance, *Space, Time and Resurrection*, p. 128.
41   Torrance, *Space, Time and Resurrection*, p. 129.
42   Torrance, *Space, Time and Resurrection*, p. 129f.
43   Torrance, *Space, Time and Resurrection*, p. 134.

God, Space, and Spatiality-Case Studies

limits of man's existence on earth) but are *infinitely open at God's end*, open to God's own eternal Being and the infinite room of his divine life."[44] It is through the Holy Spirit that, Torrance asserts, "Christ is nearer to us than we are to ourselves, and we who live and dwell on earth are yet made to sit with Christ 'in heavenly place,' partaking of the divine nature in him."[45]

Space and time are not receptacles apart from bodies or forces, but are functions of events in the universe and forms of their orderly sequence and structure. They are "relational and variational concepts defined in accordance with the nature of the force that gives them their field of determination."[46] Torrance admits there is a difference between space and time: "whereas space is three-dimensional, time is one directional or irreversible."[47] But, he contends that "in the nature of the case we cannot separate space from time, or location from time – temporal relation belongs to location."[48] In modern thought we think of space-time in a four dimensional continuum. "[W]e must think of place as well as time in terms of that *for which* they exist or function," Torrance says. Regarding humanity's place and God's place, he states: "Man's 'place' is defined by the nature and activity of man as the room which he makes for himself in his life and movement, and God's 'place' is defined by the nature and activity of God as the room for the life and activity of God as God."[49] However, Torrance sets limits on the discussion on the relation-

---

44   Torrance, *Space, Time and Resurrection*, p. 131f.

45   Torrance, *Space, Time and Resurrection*, p. 135.

46   Torrance, *Space, Time and Resurrection*, p. 130.

47   Torrance, *Space, Time and Resurrection*, p. 131. Moltmann also points out the difference between time and space as follows: "Time and space are complementary in the space-time continuum of physics, but for our human experience they are not symmetrical: we can experience different times in the same space, but not different spaces at the same time" [See Moltmann, *Experiences in Theology* (Minneapolis: Fortress Press, 2000), p. 315].

48   Torrance, *Space, Time and Resurrection*, p. 130f. In this sense the view of Torrance is different from that of Pannenberg. The latter holds as follows: "The concept of time proves to be basic in this regard, for it is constitutive for that of space. The simultaneity of what is different constitutes space. Somewhat epigrammatically Georg Picht could thus say that truly 'space is time.' The implied reduction of space to time is a presupposition for a theological interpretation of the presence of God in space as the dynamic operation of the divine Spirit" [See Pannenberg, *Systematic Theology*, vol. 2 (Grand Rapids: Eerdmans, 1994), p. 90].

49   Torrance, *Space, Time and Resurrection*, p. 131.

90   *Transcendence and Spatiality of the Triune Creator*

ship between God and space-time as follows: "We do not speak of space-time in relation to God, but we must speak of the 'place' and 'time' of God in terms of his own eternal life and his eternal purpose in the divine love."[50] By this saying, he suggests the limits and possibilities of the discussions on divine spatiality as well as temporality.

2.3 God's Temporality

Space and time cannot be separated from each other. We should think of space-time in a four-dimensional continuum. But Torrance presents the reason why we split space and time: "because our eyes cannot keep track with the speed of light in the physical universe, and thereby create problems which we try to solve through artificial devices."[51] In some places, he deploys his view on the "time" of God apart from his "space" or "place."

Torrance describes Christ events, such as resurrection and ascension, in relation to time: "The resurrection of the man Jesus, and his exaltation to the right hand of the Father, mean the taking up of *human time* into God. In Christ the life of human being is wedded to eternal life. The ascension also means that this time of the new creation in Christ is hidden from us, and, as it were, held back until in the mercy of God Jesus Christ comes again to judge and renew all his creation."[52] In several passages, Torrance maintains that God is a temporal being. Quoting Athanasius, one of his theological heroes, he argues as follows: "One of the most significant was highlighted by Athanasius when he pointed out that while God was always Father he was not always Creator – there was a 'time' when God was not Creator. This implies that when God created the world out of nothing we must think of that as something completely

---

50   Torrance, *Space, Time and Resurrection*, p. 131. He continues to say that "'Time' for God himself can only be defined by the uncreated and creative life of God, and 'place' for God can only be defined by the communion of the Persons in the Divine life – that is why doctrinally we speak of the 'perichoresis' (from *chora* meaning space or room) or mutual indwelling of the Father, Son and Holy Spirit in the Trinity of God" (ibid.).

51   Torrance, *Space, Time and Resurrection*, p. 153.

52   Torrance, *Space, Time and Resurrection*, p. 98.

God, Space, and Spatiality-Case Studies

new, *new even for God!*"[53] Creation was once a new event even for God. Elsewhere Torrance states, "If God was not always Creator, the creation of the universe as reality 'external to God' was something new in the eternal Life of God."[54]

For Torrance, the incarnation of Jesus as well as the creation of the world is a new event for God: "[T]he Son of God was not always incarnate, which meant that the incarnation also was something *new even for God*. In this event we must think of there being a 'before' and an 'after' in the life of God. We cannot really grasp what that means, but it does imply that somehow 'time' characterizes the life of God."[55] These doctrines of the absolute origin of the creation and the doctrine of the incarnation, that is, the ideas that something entirely new took place, new even for God, were undoubtedly offensive to the Greek mind. Thus Torrance asserts that "[t]ogether creation and incarnation conflicted sharply with Greek philosophical categories of the necessity, immobility and impassibility of God."[56]

Since words like "was," "before," "when" and "beginning" are time-related, Torrance holds, they present us with problems when we speak of God. He proposes the reason as follows: "the time-relations they imply may not be read back into God."[57] This contention of Torrance seems to contradict his own view above: "we must think of there being a 'before' and an 'after' in the life of God [...] somehow 'time' characterizes the life of God."[58] He presents two ways of solution. When these time-related terms are used of God and of creatures, they have a different meaning. "These terms have one sense when used of God when they are governed by the unique nature of God," Torrance writes, "and another sense when used of creatures in accordance with their transitory na-

---

53 Torrance, *Preaching Christ Today*, p. 69.
54 Torrance, *The Trinitarian Faith: The Evangelical Theology of the Ancient Catholic Church* (Edinburgh: T&T Clark, 1988), p. 88.
55 Torrance, *Preaching Christ Today*, p. 69. But he does not stop here. Torrance wants to say, "Just as the creation was a new event, new even for God, for while God was always Father, he was not always Creator; and just as the incarnation was a new event, new even for God, for the Son of God was not always incarnate; so the atonement was a new act, new even for God" [Torrance, *The Mediation of Christ* (Colorado Springs: Helmers & Howard, 1992), p. 114].
56 Torrance, *The Trinitarian Faith*, p. 89.
57 Torrance, *The Trinitarian Faith*, p. 87f.
58 Torrance, *Preaching Christ Today*, p. 69.

92 *Transcendence and Spatiality of the Triune Creator*

tures."[59] Also he proposes that there are two kinds of time. Torrance argues, "We distinguish between the uncreated light of God and created light, between the uncreated life of God and created life, just as we distinguish between uncreated being and created being. Similarly I believe we must think of uncreated time as well as created time."[60] In this connection, he refers to "what Einstein called the 'relativity of simultaneity,' according to which it is possible for an event in the physical world to have two different 'real times.'"[61]

2.4 Critical Evaluation

Along with Einstein, Michael Polanyi (1891–1976) has had more influence on Torrance's thought than any other philosopher of science. The tacit dimension is central to Polanyi's position. Colyer comments on it: "the tacit dimension, in general and in any field of scientific inquiry, is informal, undefined and in large measure inarticulate, yet still of critical significance in explicit thought and conceptual formulation."[62] Thus it implies that "in our knowledge of God through Christ and in the Spirit we know more than we can bring to explicit articulation, and that the reality of God is greater than all of our theological formulations."[63]

Polanyi and Torrance, invoking the tacit dimension, emphasize the point of indwelling the field of inquiry. As Colyer explains, "Indwelling is a holistic, significantly informal, integrative and heuristic process of investigating a field of inquiry. As this inquiry develops, our minds begin to assimilate the internal constitutive relations embodied in what we seek to know."[64] Thus adopting Polanyi's and Torrance's perspective, Colyer asserts, "we 'indwell' the corpus of Torrance's writings until we gain an insight into the intrinsic structure, the internal relations of his theology, which we then test and refine through our continued research.

---

59 Torrance, *The Trinitarian Faith*, p. 87f.
60 Torrance, *Preaching Christ Today*, p. 69.
61 Torrance, *Preaching Christ Today*, p. 70.
62 Colyer, *How to Read T. F. Torrance*, p. 335. Torrance states that "We know more than we can tell" and that "We know more physics by the age of five than we will be able to understand even if we became great scientists" (ibid.).
63 Colyer, *How to Read T. F. Torrance*, p. 336, note 64.
64 Colyer, *How to Read T. F. Torrance*, p. 336.

God, Space, and Spatiality-Case Studies 93

This entails an irreducible element of creative imagination, but an imagination controlled from beyond by the field under investigation – Torrance's theology in his publications."[65]

In order to understand one person's thought rightly and to criticize it correctly, first of all, we should look at the tacit dimension of the thought and then indwell the corpus of the person's writings. Before we gain an insight into its intrinsic structure, we may be tempted to offer some criticism of it. But then we are liable to make a mistake. We should keep in mind Torrance's view on indwelling as well as the tacit dimension. It is difficult to understand other person's thought accurately. In some sense, it does not seem to be possible. Especially it is still more difficult to understand the scholars such as Torrance whose writings are hard and sentences are long.

For me, Torrance does clearly present two problematic points. As cited above, Torrance contends, "we must think of there being a 'before' and an 'after' in the life of God."[66] However, in a different context, he asserts that "Their [Son and Spirit] being of the Father and with the Father is beyond beginning and beyond time (ἀανάρχως και ἀχρόνως) – there is no 'before' or 'after' in God. 'There never was when the Spirit was not.'"[67] Elsewhere he holds again that "God the Father, who is himself without generation or origination, is the one Principle or Origin (ἀρχή) and Cause (αἰτία) of the Son and the Spirit, although in such a way that there is no interval of existence, time or space between them and no 'before' or 'after' in the order of their being."[68]

As already cited above, Torrance writes, "[w]e do not speak of spacetime in relation to God, but we must speak of the 'place' and 'time' of God in terms of his own eternal life and his eternal purpose in the divine love."[69] In several passages, Torrance speaks of the time of God. He seems to opt for a kind of divine temporality. Torrance interprets creation and incarnation, and even atonement as something new even for God. Thus, he states that somehow "time" characterizes the life of God. But Torrance does not refer to God's spatiality except for "perichoresis" as a place for God: "'Time' for God himself can only be defined by the

---

65   Colyer, *How to Read T. F. Torrance*, p. 337.
66   Torrance, *Preaching Christ Today*, p. 69.
67   Torrance, *The Trinitarian Faith*, p. 222.
68   Torrance, *The Trinitarian Faith*, p. 237.
69   Torrance, *Space, Time and Resurrection*, p. 131.

94 *Transcendence and Spatiality of the Triune Creator*

uncreated and creative life of God, and 'place' for God can only be defined by the communion of the Persons in the Divine life – that is why doctrinally we speak of the *'perichoresis'* (from *chora* meaning space or room) or mutual indwelling of the Father, Son and Holy Spirit in the Trinity of God."[70] To my regret, he does not go further than this point. Even though he presents many useful suggestions for a discussion on the relation between God and his spatiality, Torrance does not develop his thoughts fully.

Hans Schwarz develops his own view on God's spatiality from Torrance's idea of the multi-levelled structure of human knowledge. In his article, "God's Place in a Space Age," Schwarz proposes the metaphor of a dimensional relationship between God and the world. According to him, his model makes sense for both divine transcendence and immanence alike: "God would be present in our dimension (i.e., four-dimensional space-time continuum) without being contained by it and he would transcend it without being absent."[71] Torrance's thought of redemption of space also needs further development. From that idea we can infer God's own space different from that of creatures. Torrance definitely defends a view of divine temporality. He acknowledges that God has his own time which is different from that of creatures. In parallel to that, I think we have the right to expect a more developed idea regarding the spatiality of God from Torrance. For Torrance, according to Mark Worthing, "any attempt to explicate knowledge of God apart from the structure of space-time which God created is inevitably irrational."[72] If so, Torrance should have developed divine spatiality in addition to divine temporality.

Torrance is "a theologian enamoured of the field concept" with Pannenberg.[73] With a view of the universe as "a unitary open system," Torrance claims, Einstein attempted to construct a unified field theory. Torrance states, "There now emerged the concept of the continuous field of space-time which interacts with the constituent matter/energy of the universe, integrating everything within it in accordance with its unitary yet

70 Torrance, *Space, Time and Resurrection*, p. 131.
71 Hans Schwarz, "God's Place in a Space Age," *Zygon*, 21. 3 (September 1986), p. 367.
72 Worthing, *God, Creation, and Contemporary Physics*, p. 5.
73 Polkinghorne, *Reason and Reality: The Relationship between Science and Theology* (Valley Forge, PA: Trinity Press International, 1991), p. 93.

God, Space, and Spatiality-Case Studies 95

variable, objective rational order of non-causal relations."[74] In this respect, Pannenberg has noted that Torrance is the first person to have recognized the relation between the Spirit and the field: "We must credit T. F. Torrance with being the first to recognize these links and plead for theology's adoption of the field concept."[75]

3. Pannenberg-Spirit as a Force Field

This section treats Pannenberg's view on Spirit as a force field. Subsection 1, below, introduces the basic idea of Pannenberg's explanation on the activity of the Spirit as a field. He presents the pneuma concept of the Stoics – which was virtually a material reality – as the predecessor of the field concept of modern physics. But since Origen, this view has been rejected, and Pannenberg himself definitely opposes the Stoic concept of the pneuma. Whether Pannenberg uses the terms "field" and "force" as metaphors or not is controversial. Subsection 2 examines Pannenberg's conception of time and space. He asserts divine spatiality as well as divine temporality. He claims that the idea of God's transcendence does not contradict divine spatiality. Rather, transcendence demands spatiality. But he thinks of time as more fundamental than space. Subsection 3 deals with Pannenberg's view of God's omnipresence. He argues that divine omnipresence denotes a sacramental view of nature. This sacramental view leads to a veneration of nature. Finally, subsection 4 evaluates Pannenberg's thought, arguing that Pannenberg has offered helpful reflection on God's activity, or immanence, within the world of space and time, his concept of the Spirit as a force field is a panentheistic idea.

---

74   Torrance, "Divine and Contingent Order," A. R. Peacocke ed., *The Sciences and Theology in the Twentieth Century* (University of Notre Dame Press, 1981), p. 92.

75   Pannenberg, *Systematic Theology* 2, tr. G. Bromiley (Grand Rapids: Eerdmans, 1994), p. 82, note 212. Here Pannenberg quotes Torrance: "The field that we are concerned with is surely the interaction of God with history understood from the axis of Creation-Incarnation [...] Our understanding of this field will be determined by the force or energy that constitutes it, the Holy and Creator Spirit of God" (Torrance, *Space, Time and Incarnation*, p. 71).

96   *Transcendence and Spatiality of the Triune Creator*

For, Pannenberg clearly uses the field theory as more than mere metaphor.

3.1 Spirit as a Field

Pannenberg argues that "the Spirit of God is the life-giving principle, to which all creatures owe life, movement, and activity [...] Conversely, all life perishes when God withdraws his Spirit (Ps. 104:29; Job 34:14f.). The souls of all living things and the breath of all people are in the hands of the Spirit (Job 12:10)."[76] For Pannenberg, the Spirit is "the principle of the active presence of the transcendent God with his creation" and "the medium of the participation of created life in the trinitarian divine life."[77] Especially in explaining the former principle, Pannenberg adopts the concept of the field in modern physics. According to him, "The Spirit is the force field of God's mighty presence (Ps. 139:7). This unique understanding of the Spirit of God leads on to what Ps. 139 says about God's all-embracing knowledge, which rests indeed on God's presence with all his creatures."[78] Pannenberg also says that "The Spirit of God can be understood as the supreme field of power that pervades all of creation. Each finite event or being is to be considered as a special manifestation of that field, and their movements are responsive to its forces."[79]

Concerning the reasons for introducing the field concept into theology, Pannenberg refers to his doctrine of God in which he treats the traditional description of God as Spirit alongside God as love: "Criticism of this traditional way of speaking about God as though the reference were to subjectivity *(nous)* led us to the insight (vol. I, 372ff.) that it is more in keeping with what the Bible says about God as Spirit, or about the Spirit of God, to view what is meant as a dynamic field that is structured in trinitarian fashion, so that the person of the Holy Spirit is one of the per-

---

76   Pannenberg, *Systematic Theology* 2, p. 76f. Hereafter "*ST* 2."
77   Cf. Christoph Schwöbel, "Rational Theology in Trinitarian Perspective: Wolfhart Pannenberg's Systematic Theology," *Journal of Theological Studies* 47 (1996), p. 509.
78   Pannenberg, *Systematic Theology* 1, trans. G. Bromiley (Grand Rapids: Eerdmans, 1991), p. 382. Hereafter "*ST* 1."
79   Pannenberg, "The Doctrine of Creation and Modern Science," in *Cosmos as Creation*, ed. Ted Peters (Nashville: Abingdon Press, 1989), p. 46.

God, Space, and Spatiality-Case Studies

sonal concretions of the essence of God as Spirit in distinction from the Father and the Son."[80]

Pannenberg presents the pneuma concept of the Stoics as the predecessor of the field concept of modern physics.[81] However, there is a difference between the pneuma of Stoic philosophy and the modern field concept. "For the Stoics the *pneuma* was a very fine stuff that permeates all things, that holds all things in the cosmos together by its tension *(tonos)*, and that gives rise to the different qualities and movements of things."[82] For them, the pneuma was a material element, though the finest one. On the other hand, according to Pannenberg, "the modern field concept suggests the idea of dynamic movement, of force, together with spatial and temporal extension, but without requiring a material element serving as a medium of field dynamics."[83] Nevertheless, the pneuma concept of the Stoics "influenced not only the thought of Philo but also what early Christian theology had to say about the working of the divine Spirit in creation."[84] In the later fathers, especially after Origen, however, there were fewer contacts with Stoic philosophy: "The polemic of Origen against the use of the Stoic idea of pneuma in Christian theology had been that God cannot be imagined as a material body or element, since then the deity would be composed of parts."[85] Thus Origen identified pneuma with intelligence. Pannenberg presents the fateful effect of Origen's identification as follows: "the relation of the divine Spirit to the material world and to the process of its creation was obscured. In addition, the divine Spirit was also separated from the created spirit, the human soul."[86] Pannenberg responds to Origen's criticism of the Stoic doctrine of the material nature of the pneuma, by saying this:

> But now we can see that the modern field concept comes much closer to the biblical description of God as spirit, without involving the concept of God in the prob-

---

80 *ST* 2, p. 83.

81 Pannenberg owes this part to Max Jammer: "Max Jammer thinks that the Stoic doctrine of the divine *pneuma* was actually the direct precursor of the modern field concept" (*ST* 2, p. 81).

82 *ST* 2, p. 81.

83 Pannenberg, "Theology and Science" *Princeton Seminary Bulletin* 13/3 (1992), p. 307.

84 *ST* 2, p. 81.

85 Pannenberg, "Theology and Science," p. 307.

86 Pannenberg, "The Doctrine of Creation," p. 44.

98 *Transcendence and Spatiality of the Triune Creator*

lems of bodily existence. We may imagine, then, the reality of God in terms of the *comprehensive field of eternity*, comprising time and space and dynamically producing the temporal existence of creatures in space through its futurity in relation to all potential events.[87]

In any case, Pannenberg does not maintain that the Spirit *is* such a field; instead, the Spirit is "a unique manifestation (singularity) of the field of the divine essentiality."[88] He thus seeks to use the field concept analogically to explain the divine creative presence in the universe: "The field theories of science, then, can be considered as *approximations* to the metaphysical reality of the all-pervading spiritual field of God's creative presence in the universe."[89] Therefore, Pannenberg differentiates the theological use of the field from its scientific use: "In the working of the Spirit, as in that of the divine Logos, the future of the consummation in the kingdom of God predominates. Theological talk about the dynamic of the Spirit of God in creation differs in this regard from the field theories of physics that work in terms of natural laws."[90]

Pannenberg extends this new pneumatology of field to the doctrine of God and the anthropology. Pannenberg rejects the classical understanding of God as reason and will. For him, this conception of God is a mere projection in a Feuerbachan sense. Pannenberg understands the divine essence in terms of the "incomprehensible field" and "the comprehensive field." He states, "the presence of God's Spirit in his creation can be described as a field of creative presence, a comprehensive field of force that releases event after event into finite existence."[91] For Pannenberg, thus Grenz and Olson contend, "God [...] is the 'field' in which creation

87 Pannenberg, "Theology and Science," p. 307.
88 *ST* 2, p. 83; cf. pp. 105, 109, 110. As I will note later, however, Pannenberg's contention of Spirit as a field is controversial. Many scholars, such as Ted Peters, Mark Worthing, and John Polkinghorne, except Pannenberg himself, do not agree that his use of the Spirit as a field is not just a metaphor. For the details, see subsection 4 of this section.
89 Pannenberg, "The Doctrine of Creation," p. 47.
90 *ST* 2, p. 109. Pannenberg presents the usefulness of the field concept in another way: "The turn toward the field concept in the development of modern physics has theological significance. This is suggested not only by its opposition toward the tendency to reducing the concept of force to bodies or masses, but also because field theories from Faraday to Einstein claim a priority for the whole over the parts" (Pannenberg, "The Doctrine of Creation and Modern Science," p. 164).
91 Pannenberg, *An Introduction to Systematic Theology*, p. 49.

God, Space, and Spatiality-Case Studies 99

and history exist."[92] God is in some sense "the environmental network" or "field" in which and from which creatures live. For Pannenberg, human person is not to be seen in terms of an "I" who preexists experience of the world. The immediate perception of the totality of a person's existence is important for his/her identity development. Pannenberg terms this perception "feeling" or the "field" in which a person lives.[93]

One of the crucial criticisms to Pannenberg's view of God as the divine field working in the world is the question of God's personhood. Pannenberg does not like traditional notions of God as mind and will. And the language of "field" which he adopts, implies an impersonal or suprapersonal God. This is "a God who is the whole that is greater than the sum of the world's parts but not a gracious, completely free and self-sufficient divine person."[94] Concerning other problems of Pannenberg's idea of spirit as a field, after examining some related thoughts of Pannenberg, I will deploy in detail.

3.2 Time and Space

Pannenberg links the dynamic of the Spirit as a field to time and space: "We are thus to think of the dynamic of the divine Spirit as a working field linked to time and space – to time by the power of the future that gives creatures their own present and duration, and to space by the simultaneity of creatures in their duration."[95] According to him, "The theological use of the field concept in describing God's creative presence and activity in the world of creation does require, however, a theological interpretation of space and time."[96]

Pannenberg insists that the spatiality of God is inevitable. Concerning the relation between God and space, therefore, Pannenberg observes,

---

92 Grenz and Olson, *20th-Century Theology*, p. 194.
93 Cf. Grenz and Olson, *20th-Century Theology*, p. 194. They refers to Pannenberg, *Anthropology in Theological Perspective* (Philadelphia: Westminster, 1985), pp. 226–29, 235–36, 240, 384 and Pannenberg, "Spirit and Mind," in *Mind in Nature*, ed., Richard Q. Elvee, Nobel Conference 17 (New York: Harper and Row, 1982), pp. 137, 143.
94 Grenz and Olson, *20th-Century Theology*, p. 199.
95 *ST* 2, p. 102.
96 Pannenberg, "The Doctrine of Creation," p. 47.

100   *Transcendence and Spatiality of the Triune Creator*

"Spatial ideas occur elsewhere in biblical statements about God's relation to creation. We read of God's dwelling in heaven. From there his power, or he himself, is manifested on earth (vol. I, 410ff.). The idea of God's transcendence also demands space if it is not to be reduced to the logical distinction of the Infinite from everything finite."[97] "In creating," he asserts, "God gives creatures space alongside himself and over against himself. But his presence still comprehends them. As the early fathers said, God comprehends all things and is comprehended by nothing and no one."[98]

In several places Pannenberg mentions the fact that "the great Isaac Newton conceived of space as the medium of God's omnipresence" or "as a medium of God's creative presence at the finite place of his creatures in creating them."[99] For this idea, Leibniz accused Newton of pantheism. For Newton seems to have treated space as an attribute of God himself. Samuel Clarke, however, defended Newton against that criticism "by declaring that God of course is not composed of spatial parts, but that the divine immensity is equivalent to undivided infinite space." "According to Clark," Pannenberg says, "the expression *sensorium Dei* does not mean that God needs space but that he uses it as a means for the creation of creatures each in its own place." Concerning Einstein's criticism of Newton's view of absolute space, Pannenberg contends that Einstein did not invalidate the idea of Newton, "since Einstein is not to be understood as merely opposed to Newton, but extended the function of the concept of space in Newton into a general field theory of space-time."[100] Pannenberg, however, rightly indicates that Newton was unable satisfactorily to explain the union of transcendence and presence in God's relation to his creature, because of the lack of the trinitarian scheme in his system.[101]

---

97   *ST* 2, p. 85.

98   *ST* 2, p. 86.

99   Pannenberg, "Theology and Science," p. 305. See also Pannenberg, "The Doctrine of Creation and Modern Science," p. 168; and Pannenberg, "The Doctrine of Creation," p. 48.

100 *ST* 1, p. 413f.

101 *ST* 1, p. 414. He also asserts that, "Only the doctrine of the Trinity could basically clarify the question of union and tension between transcendence and immanence (ibid., p. 415).

God, Space, and Spatiality-Case Studies 101

Time and space, for Pannenberg, are not different entities: "They belong together in the description of natural processes." Yet, he maintains that time is the more fundamental reality. Pannenberg thus asserts the following:

> The concept of time proves to be basic in this regard, for it is constitutive for that of space. The simultaneity of what is different constitutes space. Somewhat epigrammatically Georg Picht could thus say that truly "space is time." The implied reduction of space to time is a presupposition for a theological interpretation of the presence of God in space as the dynamic operation of the divine Spirit.[102]

As the concept of time depends upon eternity, therefore, the concept of space depends upon the immensity and omnipresence of God.[103] Thus Pannenberg argues that "God's immensity and eternity can be regarded as constitutive of time and space, and consequently it makes sense to speak of a field of God's spiritual presence in his creation."[104]

This thought of Pannenberg on space does not seem to overcome the limitation of Newton's view of absolute space. Here is Torrance's explanation about Newton's view: "Newton himself spoke of space and time as an infinite receptacle in terms of the infinity and eternity of God, for it is in God as in a container that we live and move and have our being. Thus infinite volume is related in his thought to the Spirit of God and infinite time is identical with eternity – infinite space and time are in fact attributes of Deity."[105] In this sense, Pannenberg would have had the relational, relative, or differential concept of space.[106]

---

102 *ST* 2, p. 90.
103 Pannenberg, "Theology and Science," p. 306.
104 Pannenberg, "The Doctrine of Creation," p. 48 f.
105 Torrance, *Space, Time and Incarnation*, p. 38.
106 Cf. Torrance, *Space, Time and Incarnation*, p. 18, explaining the thought of Athanasius who is the Nicene theologian, states: "Space is here a *differential* concept that is essentially *open-ended*, for it is defined in accordance with the interaction between God and man, eternal and contingent happening."

102   *Transcendence and Spatiality of the Triune Creator*

3.3 Omnipresence

Pannenberg avers, "God's presence permeates and comprehends all things."[107] God incommensurably transcends his creation, but at the same time he is still present to even the least of his creatures.[108] Thus, Pannenberg writes, "In order to be omnipresent, God cannot be merely transcendent in relation to the world of space, but must be conceived of as also present in each part of space."[109]

Pannenberg connects omnipotence and omnipresence with the eternity of God, and interprets divine omnipresence as the dynamic of the Spirit: "Omnipotence and omnipresence are very closely related and are also closely related to God's eternity. As all things are present to God in his eternity, and he is present to them, so he has power over all things. His omnipresence for its part is full of the dynamic of his Spirit."[110]

This explanation of God's omnipresence, for Pannenberg, connotes "a sacramental view of nature." He writes:

> Oriented to the eschatological consummation, this creative presence of God in the immanence of the creaturely world and its forms has given us occasion today to speak of a "sacramental" view of nature in Christianity. Like the matter of the sacraments, the material universe is not just an outwardly visible sign of an invisible grace (God's presence) but a means to communicate it.[111]

However, according to Pannenberg, this sacramental view has a tendency to slip back into veneration of nature. "General references to the sacramental nature of the universe," he says, "generally omit the eschatological reference that is decisive for the understanding of nature in Rom. 8:19ff." Thus Pannenberg contends that "As creatures of God, constructs in the world of nature are both less than this and more – less

---

107   *ST* 1, p. 411.
108   *ST* 1, p. 412.
109   Pannenberg, "Theology and Science," p. 305.
110   *ST* 1, p. 415.
111   *ST* 2, p. 137. Cf. Torrance, *Theological Science* (Edinburgh: T&T Clark, 1996), p. 66, states, "In the Augustinian tradition which dominated the Middle Ages, the universe was regarded as a sacramental macrocosm in which the physical and visible creation was held to be the counterpart in time to eternal and heavenly patterns. Thus the world of nature was looked at only sacramentally, i.e. looked through toward God and the eternal realities." In this view of nature, according to Torrance, the development of empirical science would not be expected.

God, Space, and Spatiality-Case Studies 103

because they do not have the specific orientation to the presence of future eschatological salvation in Jesus Christ that is bound up with the institution of a sacrament, more because natural forms themselves are the object of the divine will in creation and do not have their meaning in orientation to something else." [112] Hence he maintains that any talk of the sacramental reality of creaturely things is too general and imprecise. Nevertheless, Pannenberg positively evaluates "the thought that Teilhard de Chardin emphasizes, namely, that in the sacraments of the new covenant, and above all in the eucharistic bread and wine, all creation is taken up into the sacramental action of thanksgiving to God." [113]

3.4 Critical Evaluation

Although he does not agree with Pannenberg's position, Colin E. Gunton admits there is merit to field theory in one sense: "Whereas mechanism suggests a world impervious to anything but the laws of its own being, except in terms of miracles which 'violate' supposed 'laws' of nature, field theory at least enables it to be conceivable that nature may be generally law-abiding, but contingently rather than rigidly so." [114] According to Gunton, Pannenberg's explanation of the Holy Spirit's economic activity in terms of fields of force is an attempt to think of physical reality as interacting fields of force rather than billiard-ball-like entities bumping into one another. This attempt is extremely important, Gunton contends, since it can suggest that the world is open to God's continuing interaction with it.

Ted Peters also indicates some theological significances of Pannenberg's explanation of the Spirit as a field. One of the results of the post-Newtonian reduction of forces to mass in motion is the preclusion of any divine force. Thus, in a post-Newtonian universe, Peters contends, "If God does not have a body, and if all forces require a prior body, then God cannot have force. This problem is eliminated with contemporary field theory." [115] The field theory claims a priority for the whole over the

112 *ST* 2, p. 138.
113 *ST* 2, p. 138.
114 Gunton, *The Triune Creator*, p. 175.
115 Ted Peters, "Pannenberg on Theology and Natural Science," in *Toward a Theology of Nature*, ed. T. Peters (Westminster/John Knox Press: Louisville, 1993), p. 13.

104   *Transcendence and Spatiality of the Triune Creator*

parts and provides a possible means for conceiving the divine Spirit as active in the natural world. Even more importantly "Pannenberg employs the notion of a dynamic field to describe the workings of the Spirit within the Trinitarian life proper."[116] Peters, however, does not overlook the danger of this thought: "Pannenberg rushes in where two-language angels have feared to tread. He does not say that spirit *is like* a force field. He says spirit *is* a force field [...] Historians of science are quick to point out the dangers of trying to float a theological assertion aboard a scientific ship, because the intellectual weather can change suddenly."[117]

It is debatable whether Pannenberg uses words such as "field" and "force" merely in an analogical sense. Mark W. Worthing notes, "The analogical character of the concepts of force and fields of force is significant for the discussion of the implications of such concepts for theology."[118] But, John Polkinghorne warns against going beyond the use of fields of force as analogical models and identifying them with God or God's activity. He argues, "To take it more seriously than that would be to regard quantum fields as the 'sensorium of God' (in the way that Newton was emboldened to talk about absolute space), a panentheistic idea which does not commend itself to me. When all is said and done, quantum fields are simply creatures."[119]

Concerning Pannenberg's use of field theory, Worthing points out that Pannenberg insists that he does not intend to reduce the Holy Spirit to phenomena, accounted for by field theory but proposes only a model. Worthing quotes the following passage of Pannenberg:

> To be sure, even a cosmic field conceived along the lines of Faraday's thought as a field of force would not be identified immediately with the dynamic activity of the divine Spirit in creation. In every case the different models of science remain approximations [...] Therefore, theological assertions of field structure of the cosmic activity of the divine Spirit will remain different from field theories in physics.[120]

Pannenberg differentiates between physical, philosophical, and theological uses of the field concept. However, in seeking to develop new con-

---

116   Peters, "Pannenberg on Theology and Natural Science," p. 14.
117   Peters, "Pannenberg on Theology and Natural Science," p. 14.
118   Worthing, *God, Creation, and Contemporary Physics*, p. 119.
119   Worthing, *God, Creation, and Contemporary Physics*, p. 120.
120   Pannenberg, "The Doctrine of Creation," p. 14, cited by Worthing, *God, Creation, and Contemporary Physics*, p. 120f.

God, Space, and Spatiality-Case Studies 105

ceptual models for the Trinity, the Holy Spirit, and angels, in light of the concept of field theory, Worthing says, "as in his doctrine of angels, he [Pannenberg] seems to identify these messengers of God with specific physical forces and fields and thus confuse the theological and physical concepts of field."[121] According to Worthing, this difficulty can also be seen in Pannenberg's doctrine of God.

In his doctrine of God, Pannenberg clearly makes a distinction between theological and scientific concepts of field. He states, "The person of the Holy Spirit is not itself to be understood as field but rather as a unique manifestation (singularity) of the field of the divine essence. But because the person of the Holy Spirit is only revealed in relation to the Son (and thus also the Father), its action in the creation has more the character of the dynamic working of a field."[122] However, "when he makes the unification of space and time as contained in relativity theory a precondition for a theological interpretation of the 'dynamic presence of the Holy Spirit,'" Worthing points out, "Pannenberg seems to link his conceptual model of God as Spirit to a specific viewpoint."[123] Pannenberg claims that the "reduction of space to time is a prerequisite for theological interpretation of God's presence in space as the dynamic working of the divine Spirit."[124] By saying this, Worthing contends, "Pannenberg seeks to incorporate Einstein's field theory of an integrated space-time into his understanding of God's continuing dynamic sustenance of the world through the Holy Spirit – which is a part of the larger trinitarian field that Pannenberg identifies as God."[125]

As Worthing thus observes, Pannenberg integrates angels into his understanding of field. Pannenberg states that, "From the point of view of the field structure of spiritual dynamics one could consider identifying the subject matter intended in the conception of angels with the emergence of relatively independent parts of the cosmic field."[126] Even further, Pannenberg identifies some of the particular forces and dynamic spheres with angels or demons: "If such forces as Wind, Fire, and Stars are identified as angels of God then they would be (explained) themati-

---

121 Worthing, *God, Creation, and Contemporary Physics*, p. 121.
122 *ST* 2, p. 104.
123 Worthing, *God, Creation, and Contemporary Physics*, p. 121f.
124 *ST* 2, p. 111.
125 Worthing, *God, Creation, and Contemporary Physics*, p. 122.
126 Pannenberg, *Toward a Theology of Nature*, p. 41.

106 *Transcendence and Spatiality of the Triune Creator*

cally in their relationship to God as creatures and likewise in view of the mixed experiences of them by human beings who experience them either as servants of God or as demonic powers which strive against the will of God."[127] On these notes, Worthing raises important questions as follows:

> Is there ultimately a qualitative distinction between Pannenberg's proposal to understand angels as physical force fields and identifying God with the sum total of physical laws or forces in the universe? That is to say, is there a danger in Pannenberg's program that angels, the Holy Spirit, and ultimately God could be reduced to metaphors for physical realities? John Polkinghorne points to a similar difficulty when he suggests that Pannenberg's use of field theory is "dangerously close to Newton's equation of absolute space with the sensorium of God."[128]

In addition to this criticism, Worthing contends that Pannenberg does not distinguish between various field theories. In the natural sciences there are a number of field theories, biological as well as physical.[129] And within physics alone there are different field theories, classical and quantum. Borrowing ideas from various and sometimes unrelated field theories, Worthing notes, Pannenberg builds his theological field model on the general idea of fields of force. It seems Pannenberg's program mostly relies on Faraday's field theory. According to Pannenberg, Faraday's vision is "to reduce all the different forces to a single field of force that determines all the changes in the natural universe."[130] Against this point, Worthing avers:

> Modern physics, however, while seeking a unified field *theory* that would combine classical and quantum field theories, has long ago abandoned the idea that a single unified *field* exists. By relying on pre-Einsteinian, pre-quantum theory concepts of field, Pannenberg would seem to neglect his goal of relating 'the field concept of

---

127 *ST* 2, p. 129.
128 Worthing, *God, Creation, and Contemporary Physics*, p. 123.
129 Cf. Ian Barbour, *Religion and Science: Historical and Contemporary Issues* (New York: HarperCollins, 1997), p. 196, explains four basic physical forces as follows: (1) the electromagnetic force responsible for light and the behavior of charged particles (Faraday, Maxwell, 1850–1860's); (2) the weak nuclear force responsible for radioactive decay; (3) the strong nuclear force that binds protons and neutrons into nuclei; and (4) the gravitational force evident in the long-distance attraction between masses (Newton, 1670's).
130 Pannenberg, *Toward a Theology of Nature*, p. 38.

God, Space, and Spatiality-Case Studies

modern physics to the Christian doctrine of the dynamic presence of the divine Spirit.'[131]

Jeffrey Wicken also accuses Pannenberg of misunderstanding and/or misusing the concept of field theory: "If we want to use the word *energy* or *field* in science-theology discourse, let us do so in some way commensurate with their understandings in physics [...] *Field* has been used in a spectrum of senses in science ranging from the specifically denotative to the connotative to the metaphorical. Pannenberg uses them all in pursuing a theology of wholeness in evolutionary process." Wicken concludes that, "although as metaphor this notion [of field] is rich for theology, taken literally it binds God needlessly to physics. Is God conceived here as a *field in physics*? If so, why the need for God at all?"[132]

After presenting some of the difficulties with Pannenberg's work on the field concept, Worthing sums up the value and limits of the use of the field theory in theology:

> This does not imply that field theories are irrelevant for understanding God's sustaining presence in the world. In fact, Pannenberg is certainly correct in his assessment of their importance in this regard. Likewise, as models of the contingency of all matter they are very fruitful. Field theory certainly has value as a metaphor for God's continuing sustenance of the universe and may well influence the way in which theology confesses this continuing "creative" presence of God. It would be a mistake, however, to build any part of our theology on a specific physical theory in such a way that the theory provides more than metaphors of meaning but becomes necessary for our theological formulations.[133]

By applying field theory, Pannenberg has sought to affirm divine spatiality. However, his argument implicitly leads to a panentheistic view of the Spirit as a field. In comparison to Pannenberg, Moltmann admits that he holds the position of panentheism.

---

131 Worthing, *God, Creation, and Contemporary Physics*, p. 123.
132 J. Wicken, "Theology and Science in the Evolving Cosmos," p. 52, cited by Worthing, *God, Creation, and Contemporary Physics*, p. 124.
133 Worthing, *God, Creation, and Contemporary Physics*, p. 124.

108 *Transcendence and Spatiality of the Triune Creator*

4. Moltmann-Creation as Divine *Zimzum*

This section deals with Moltmann's doctrines of creation and divine immanence. He holds the position that creation is the result of God's self-determination to be the creator of this world before creation. Even though he denies any necessity of creation, Moltmann contends that in some sense God needs the world (subsection 1). He interprets *creatio ex nihilo* in a peculiar way. Adopting the kabalistic doctrine of *zimzum*, Moltmann contends that God withdraws into himself to make room for creation. This room, as a result of God's self-limitation, is the *nihil* of *creatio ex nihilo* (subsection 2). The cosmic Shekinah is the single focus of the different dimensions of Moltmann's eschatological discussion. By the term "cosmic Shekinah," he seeks to express the mutual indwelling between God and creatures in the final state (subsection 3). Moltmann's main theological concern is God's immanence in the world. He admits his position is a kind of panentheism. Inevitably then his view of divine transcendence is less than adequate. In overemphasizing divine immanence, Moltmann fails to maintain a creative balance between divine transcendence and immanence (subsection 4).

4.1 Self-determination of God

Moltmann maintains that the world is created by God's free will, neither out of pre-existent matter, nor out of the divine Being itself. He connects the creation of this world to the divine determination of God to be a creator of a world: "If it [this world] is created through God's free will, and is not an emanation from God's essential nature, then the act of creation must be based on a divine resolve of the will to create. God determines that he will be the Creator of a world, before he calls creation into existence."[134] For Moltmann, therefore, the question, "what was God doing before he created the world?" is not a pointless question. Moltmann answers this question as follows: "before the creation of the world, God resolved to be its Creator in order to be glorified in his kingdom." So Moltmann contends as follows: "Between God's essential eternity and

---

134   J. Moltmann, *God in Creation*, p. 75.

God, Space, and Spatiality-Case Studies 109

the time of creation stands *God's time* for creation – the time appointed through his resolve to create."[135]

Moltmann opposes the doctrine of the creative emanation of the divine life. This idea was proffered by the Neoplatonists and in modern times has been maintained by Paul Tillich. Moltmann is critical of Tillich's thought: "By identifying the divine creativity with the divine life itself, Tillich is really abolishing God's self-differentiation from the world which he has created."[136] At the same time, however, Moltmann asserts that creation must not be thought of as "the work" of an arbitrary, capricious demiurge. So when he says that God created the world "out of freedom," he immediately adds, "out of love."[137] He objects to the formal concept of freedom which means an absolute right of disposition over property. He proposes the substantial concept of freedom. In the substantial sense, the truth of freedom is love, which means the self-communication of the good. As Moltmann states, "In this sense freedom and love are synonymous."[138]

Moltmann's view here can be illuminated by contrasting it with Karl Barth's. Barth sees God's freedom and his love as complementary. Thus Barth tries to mediate between his concept of liberty and the concept of God's goodness by defining God as "the One who loves in freedom." Following the Reformed tradition, Barth holds to divine self-sufficiency: "'God [...] could have remained satisfied with Himself and with the impassible glory and blessedness of His own inner life. But He did not do so. He elected man [...]'"[139] For Moltmann, however, Barth's position is inappropriate:

> [I]f he [God] himself determines not to be sufficient for himself (although he could be so), then there is after all a contradiction between his nature before and after this decision; and this would mean a contradiction between his nature and his revela-

---

135 Moltmann, *God in Creation*, p. 114.
136 Moltmann, *God in Creation*, p. 84.
137 Moltmann, *God in Creation*, p. 75.
138 Moltmann, *God in Creation*, p. 82. Elsewhere Moltmann asserts, "Freedom arrives at its divine truth through love. Love is a self-evident, unquestionable 'overflowing of goodness' which is therefore never open to choice at any time. We have to understand true freedom as being the self-communication of the good" (Moltmann, *The Trinity and the Kingdom*, p. 55).
139 K. Barth, *Church Dogmatics* II/2, p. 166; see also IV/2, pp. 345f., II/2, p. 10. Cf. Moltmann, *God in Creation*, p. 82, quotes this passage.

110   *Transcendence and Spatiality of the Triune Creator*

tion. The reasoning 'God could,' or 'God could have,' is inappropriate. It does not lead to an understanding of God's freedom. God's freedom can never contradict the truth which he himself is.[140]

Moltmann contends that God's freedom is his love. They are synonymous. God's love is the self-communication of the good. So Moltmann asserts,

> In his love God can choose; but he chooses only that which corresponds to his essential goodness, in order to communicate that goodness as his creation and in his creation [...] In his free love God confers his goodness: that is the work of his creation. Out of his free love he conveys and communicates his goodness: that is the work of sustaining his creation. His love is literally ecstatic love: it leads him to go out of himself and to create something which is different from himself but which none the less corresponds to him.[141]

God's resolve to create is "an essential resolve" on God's part. Moltmann maintains that "God *dis*closes himself in the *dec*ision he makes." Through this contention, Moltmann argues for the unity of will and nature in God: "God loves the world with the very same love which he eternally *is*. This does not mean that he cannot but love the world eternally; nor that he could either love it, or not."[142] Thus, concerning God's necessity and his freedom Moltmann maintains,

> If God's nature is goodness, then the freedom of his will lies in his will to goodness [...] If we lift the concept of necessity out of the context of compulsive necessity and determination by something external, then in God *necessity* and *freedom* coincide; they are what is for him axiomatic, self-evident [...] For God it is axiomatic to love freely, for he is God.[143]

---

140   Moltmann, *The Trinity and the Kingdom*, p. 53. Moltmann also argues that "... that would mean that God has two natures: describing his nature before his self-determination, we would have to say that God is in himself blessed and self-sufficient; whereas describing his nature afterwards, we would have to say that God is love – he chooses man – he is not self-sufficient" (ibid., p. 54).
141   Moltmann, *God in Creation*, p. 76.
142   Moltmann, *God in Creation*, p. 85. Elsewhere Moltmann states that "If he [God] is love, then in loving the world he is by no means 'his own prisoner'; on the contrary, in loving the world he is entirely free because he is entirely himself" (Moltmann, *The Trinity and the Kingdom*, p. 55).
143   Moltmann, *The Trinity and the Kingdom*, p. 107.

God, Space, and Spatiality-Case Studies 111

God created this world out of his free-willed or voluntary determination, not out of any necessity or arbitrariness. He freely created this world because he is love: "If God is love, then he neither will nor can be without the one who is his beloved." Thus Moltmann suggests that "In this sense God 'needs' the world and man."[144]

Christian theism holds the position that creation is solely the work of *God's free will*, as a work depending entirely on God, without any significance for God himself. God need not have created the world. There are no inner reasons and no outward compulsions for his action. So Christian theism contends that "God is self-sufficient."[145] On the other hand, according to Moltmann,

> *Christian panentheism...* started from the divine essence: Creation is a fruit of God's longing for his Other and for that Other's free response to the divine love. That is why the idea of the world is inherent in the nature of God himself from eternity [...] And if God's eternal being is love, then the divine love is also more blessed in giving than in receiving. God cannot find bliss in eternal self-love if self-lessness is part of love's very nature. God is in all eternity self-communicating love.[146]

Concerning the relationship between freedom and necessity in God, Richard Bauckham explains that "He [Moltmann] denies that the contrast between freedom of choice and necessity is real in God. God's freedom is not arbitrary choice but the inner necessity of his being, i.e. the God who is love cannot choose not to love."[147] From this position, even though in his earlier career Moltmann sticks to the traditional position that "God did not have to create something to realize himself," Bauckham indicates, Moltmann develops the view that "the creation of the world is a necessity of God's love."[148] It is in this sense that Moltmann is a panentheist. He is dissatisfied with the Barthian view that God could have remained self-sufficient but from eternity chose not to be. Moltmann argues that creation is a kind of necessity. But the necessity is the freedom of God's love and the freedom is the necessity of his love.

---

144 Moltmann, *The Trinity and the Kingdom*, p. 58.
145 Moltmann, *The Trinity and the Kingdom*, p. 105.
146 Moltmann, *The Trinity and the Kingdom*, p. 106.
147 Richard Bauckham, *The Theology of Jürgen Moltmann* (Edinburgh: T&T Clark, 1995), p. 109.
148 Bauckham, *The Theology of Jürgen Moltmann*, p. 159.

112 *Transcendence and Spatiality of the Triune Creator*

On this point, however, Moltmann's position is different from the panentheism of process theologians (i.e., "ontological panentheism"). Unlike process theologians, he denies that creation is the result of a divine inner necessity. In this sense, the difference between Moltmann's "voluntary panentheism"[149]and classical theism is not absolute. Bauckham explains this in the following way: "Moltmann maintains the priority of God's voluntary love towards the world. It is only because God voluntarily opens himself in love to the world that he can be affected by it."[150]

Steven Bouma-Prediger also correctly indicates that for Moltmann creation is in fact necessary and not contingent. He writes: "Intratrinitarian love is insufficient. God cannot exist without creation. Creation is necessary in order to complete divine love."[151] Even further, in contrast to his position that he opposes to the Neo-platonic idea of emanation in *God in Creation*, Moltmann wants to incorporate the strengths of the emanation tradition in *The Spirit of Life*: "In pneumatology we have to take up the concept of emanation, which has been falsely denigrated as neoplatonic; for through emanation, created being will be 'deified', and God is glorified in what he has created."[152]

---

149 I regard the panentheism of process theologians as "ontological panentheism." In contrast to it, Moltmann's panentheism can be dubbed "voluntary panentheism" I first applied this term to Pannenberg. In my Th. M. thesis I dub Pannenberg's view "voluntary panentheism." See "The Voluntary Panentheism of Wolfhart Pannenberg" (Th. M. Thesis, Calvin Theological Seminary, 1999). I think Moltmann is also a voluntary panentheist in that he, unlike process theologians, emphasizes the voluntary freedom of God rather than the necessity of his being in the doctrine of creation. In her recent Ph. D dissertation, Oksu Shin refers to Moltmann's panentheism as "Trinitarian, voluntary, and eschatological panentheism. See Oksu Shin, "The Panentheisic Vision in the Theology of Jürgen Moltmann" (Ph. D. Dissertation, Fuller Theological Seminary, 2002).
150 Bauckham, *The Theology of Jürgen Moltmann*, p. 58f.
151 Steven Bouma-Prediger, *The Greening of Theology: The Ecological Models of Rosemary Radford Ruether, Joshep Sittler, and Jürgen Moltmann* (Atlanta, GA: The American Academy of Religion, 1995), p. 253.
152 Moltmann, *The Spirit of Life: A Universal Affirmation* (Minneapolis: Fortress Press, 1992), p. 283.

God, Space, and Spatiality-Case Studies 113

4.2 God's Self-limitation

Prior to "the creation of a world which is not God, but which none the less corresponds to him," Moltmann assumes "a *self-limitation* of the infinite, omnipresent God." He distinguishes between "an act of God outwardly" and "an act of God inwardly" in the process of God's creation of the world. "Before God issues creatively out of himself," states Moltmann, "he acts inwardly on himself, resolving for *himself*, committing *himself*, determining *himself*."[153] Through this contention, Moltmann proposes that there is a "within" and a "without" for God. As Moltmann says, "In order to create something 'outside' himself, the infinite God must have made room for this finitude beforehand, 'in himself.'"[154] Moltmann thus presents a new understanding of the *nihil* for *God's creatio ex nihilo*: God's self-limitation is God's withdrawal into himself that can free the space in which God becomes creatively active. Through this self-withdrawal of the omnipotent and omnipresent God, the *nihil* comes into being.[155] God withdraws his presence and restricts his power.

In this context Moltmann asks whether we must not say that this "creation outside God'" exists simultaneously *in God*, in the space which God has made for it in his omnipresence. As an answer to this question, he maintains that "The trinitarian relationship of the Father, the Son and the Holy Spirit is so wide that the whole creation can find space, time and freedom in it."[156] Thus, this "creation as God's act in God and out of God," Moltmann asserts, must rather be called a feminine concept, a bringing forth: "God creates the world by letting his world become and be in *himself*: Let it be!"[157]

Moltmann develops this idea of God's self-limitation with the help of the Jewish cabbalistic doctrine of *zimsum*. "*Zimsum* means concentration

---

153 Moltmann, *God in Creation*, p. 86.

154 Moltmann, *The Trinity and the Kingdom*, p. 109.

155 See Moltmann, *The Trinity and the Kingdom*, p. 109 and Moltmann, *God in Creation*, p. 86f.

156 Moltmann, *The Trinity and the Kingdom*, p. 109.

157 Moltmann, *The Trinity and the Kingdom*, p. 109. But Moltmann does not present a view like that of Arthur Peacocke: "God creates a world that is, in principle and in origin, other than 'himself' but creates it, the world, within 'herself'" [*Intimations of Reality* (Notre Dame, IN: University of Notre Dame Press, 1984), p. 64].

114 *Transcendence and Spatiality of the Triune Creator*

and contraction, and signifies a withdrawal of oneself into oneself."[158] Moltmann writes:

> The 'existence of the universe was made possible through a shrinkage process in God.' That is his answer to the question: since God is 'all in all,' how can anything else that is not God exist at this specific point? [...] The very first act of the infinite Being was therefore not a step 'outwards' but a step 'inwards,' a 'self-withdrawal of God from himself into himself,' as Gershom Scholem puts it; that is to say, it was a *passio Dei*, not an *action* [...] It is only in Act II that God issues from himself as creator into that primal space which he had previously released in Act I.[159]

According to the doctrine of *zimsum*, the creator God is not an "unmoved mover" of the universe (i.e., the Aristotelian God). "On the contrary," states Moltmann, "creation is preceded by this self-movement on God's part, a movement which allows creation the space for its own being. God withdraws into himself in order to go out of himself."[160] Thus, according to Moltmann, this idea points to a necessary correction in the interpretation of creation: "God does not create merely by calling something into existence, or by setting something afoot. In a more profound sense he 'creates' by letting-be, by making room, and by withdrawing himself. The creative making is expressed in masculine metaphors. But the creative letting-be is better brought out through motherly categories."[161]

---

158 Moltmann, *God in Creation*, p. 87. This doctrine of *zimzum* is developed by Isaac Luria. Max Jammer indicates that Newton was under the influence of Henry More who was an ardent scholar of cabalistic lore and then connects More's conception of space to Luria's cabalistic notion of the *zimsum* (Max Jammer, *Concepts of Space*, p. 48f).

159 Moltmann, *The Trinity and the Kingdom*, p. 109f.

160 Moltmann, *God in Creation*, p. 87.

161 Moltmann, *God in Creation*, p. 88. I do not deal with the discussion about the sexuality of God in this dissertation in detail. Concerning that, see the following passages of Moltmann, *The Trinity and the Kingdom*: "The name of Father is therefore a theological term – which is to say a trinitarian one; it is not a cosmological idea or a religious-political notion [...] It is freedom that distinguishes him from the universal patriarch of father religions" (p. 163). "He is a motherly father too. He is no longer defined in unisexual, patriarchal terms but – if we allow for the metaphor of language – bisexually or transexually" (p. 164). "According to the Council of Toledo in 675 'it must be held that the Son was created, neither out of nothingness nor yet out of any substance, but that He was begotten or born out of the Father's womb *(de utero Patris)*, that is, out of his very essence'" (p. 165). "Monotheism was and is the religion of patriarchy, just as pantheism is probably the religion of earlier matriarchy. It is only the doctrine of the Trinity, with the bold

God, Space, and Spatiality-Case Studies 115

Moltmann prefers the feminine categories to the masculine ones. He develops this thought by means of the cosmic Shekinah.

4.3 *The Cosmic* Shekinah

In the preface to his book *The Coming of God*, Moltmann suggests his single focus of the different dimensions of eschatological discussion as "the cosmic Shekinah of God": "God desires to come to his 'dwelling' in his creation, the home of his identity in the world, and in it to his 'rest,' his perfected, eternal joy."[162] In God's eschatological Shekinah the whole creation will be new and eternally living, and every created thing will with unveiled face arrive at its own self. Moltmann argues,

> [T]he creation is created anew so that it may embrace 'the new Jerusalem' and become the home of God's Shekinah (Isa. 65; Ezek. 37; Rev. 21) [...] The eschatological indwelling of God in 'the new heaven and the new earth' is *the presence of God* in the *space* of his created beings. That which went up with Israel out of bondage in Egypt, that which found a temporally restricted dwelling in Jerusalem on Mount Zion – that very same presence will fill and interpenetrate the great spaces of creation, 'heaven and earth,' and will bring to all heavenly and earthly creatures eternal life and perfect justice and righteousness: God's Shekinah.[163]

With the theology of the Holy Spirit, the theology of the Shekinah is a theology of love. It is not patriarchal, but rather maternal and feministic. According to Moltmann, the Shekinah and the Holy Spirit are "the feminine principle of the Godhead."[164] By the idea of Shekinah, Moltmann seeks to express God's indwelling in creation. He states, "the divine secret of creation is the Shekinah, God's indwelling; and the purpose of

---

     statements we have quoted, which makes a first approach towards overcoming sexist language in the concept of God" (p. 165).

162 Moltmann, *The Coming of God*, p. xiii.

163 Moltmann, *The Coming of God*, p. 265f. Elsewhere Moltmann quotes the following saying of Franz Rosenzweig to explain the Shekinah: "The Shekinah, the descent of God to human beings and his dwelling among them, is conceived of as a division which takes place in God himself. God cuts himself off from himself. He gives himself away to his people. He suffers with their sufferings, he goes with them through the misery of the foreign land..." (Moltmann, *God in Creation*, p. 15).

164 Moltmann, *The Trinity and the Kingdom*, p. 57.

116   *Transcendence and Spatiality of the Triune Creator*

the Shekinah is to make the whole creation the house of God."[165] But by this term, Moltmann also attempts to express the indwelling of creation within God: "'the space of creation' is its *living space in God*. By withdrawing himself and giving his creation space, God makes himself the living space of those he has created [...] The Creator becomes the *God who can be inhabited*. God as living space of the world is a feminine metaphor, as Plato already observed."[166]

But how can the infinite God 'dwell' in earthly limited spaces and communities without destroying these spaces and communities through his infinity? Again, in response, Moltmann proposes the Jewish doctrine of the Shekinah: "The idea of the Shekinah links the infinite God with a finite, earthly space in which he desires to live. Shekinah theology is temple theology. Shekinah means the act of God's descent, and its consequence in his indwelling."[167] The concomitant idea of mutual indwelling is a definition of Shekinah theology. Moltmann understands the concept of mutual indwelling as mutual interpenetration. So, he contends, "the concept of mutual interpenetration makes it possible to preserve both the unity and the difference of what is diverse in kind: God and human being, heaven and earth, person and nature, the spiritual and the sensuous."[168]

In addition to the Jewish contention that "The Lord is the dwelling space of his world," following the exposition of Jewish and Christian Shekinah theology, Moltmann argues that creation is also destined to be the dwelling space for God. Thus he states:

> Through the historical process of indwelling and its eschatological completion, the distanced contraposition of the Creator *towards* his creation becomes the inner presence of God in his creation. To the external presence of God *above it* is added the inner presence of God *within it*. To the transcendence of the Creator towards his creation is added the immanence of his indwelling in his creation. With this the

---

165   Moltmann, *God in Creation*, pp. xiv–xv.
166   Moltmann, *The Coming of God*, p. 299.
167   Moltmann, *The Coming of God*, p. 302. He also presents *"the Christian doctrine about the incarnation of the Logos"* and "the inhabitation of the Spirit" as the answers in another form (ibid.).
168   Moltmann, *The Coming of God*, p. 278.

God, Space, and Spatiality-Case Studies 117

whole creation becomes the *house of God*, the *temple* in which God can dwell, the *home country* in which God can rest.[169]

As a result, Moltmann's frequently repeated panentheistic vision of God's being 'all in all' (I Cor. 15.28) will come true at last.[170] Let us listen to his own sayings:

> All created beings participate directly and without mediation in his indwelling glory, and in it are themselves glorified. They participate in his divine life, and in it live eternally. Once God finds his dwelling place in creation, creation loses its space outside God and attains to its place in God. Just as at the beginning the Creator made himself the living space for his creation, so at the end his new creation will be his living space. A mutual indwelling of the world in God and God in the world will come into being. For this, it is neither necessary for the world to dissolve into God, as pantheism says, nor for God to be dissolved in the world, as atheism maintains. God remains God, and the world remains creation. Through their mutual indwellings, they remain unmingled and undivided, for God lives in creation in a God-like way, and the world lives in God in a world-like way.[171]

Through such mutual indwellings of God and the world, a so-called "cosmic *communicatio idiomatum*, a communication of idioms" takes place. In order to express this idea, Moltmann uses a scholastic phrase "mutual participation in the attributes of the other":

169 Moltmann, *The Coming of God*, p. 307.
170 Moltmann, *The Trinity and the Kingdom*, p. 104f, contends as follows: "The whole world will become God's home. Through the indwelling of the Spirit, people and churches are already glorified *in the body*, now, in the present. But then the whole creation will be transfigured through the indwelling of God's glory. Consequently the hope which is kindled by the experience of the indwelling Spirit gathers in the future, with panentheistic visions. Everything ends with God's being 'all in all' (I Cor. 15.28 AV). *God in the world* and *the world in God* – that is what is meant by the glorifying of the world through the Spirit. That is *the home of the Trinity.*"
171 Moltmann, *The Coming of God*, p. 307. Similarly, Moltmann states the following with regard to relationship between the Persons of the Trinity: "The difference between Shekinah eschatology and Christian eschatology is the difference between the new temple in the earthly Jerusalem, and the new Jerusalem that comes down from heaven and which has no temple. Ultimately, in the redemption, God and his Shekinah will be indistinguishably one. God's 'self-differentiation' will be ended. According to the Christian idea, Christ will hand over the kingdom to the Father, so that God may be 'all in all,' but the Son does not therefore make himself superfluous, and there is no self-dissolution of the Son in the Father. In glory too, God is still the triune God" (ibid., p. 306).

118   *Transcendence and Spatiality of the Triune Creator*

> Created beings participate in the divine attributes of eternity and omnipresence, just as the indwelling God has participated in their limited time and their restricted space, taking them upon himself. This means that for those God has created, the time *(chronos)* of remoteness from God and of transience ceases, and eternal life in the divine life begins. It means that for those God has created, the space *(topos)* of detachment from God ceases, and eternal presence in the omnipresence of God begins. God's indwelling eternity gives to created beings eternal time. God's indwelling presence gives to created beings for ever the 'broad space in which there is no more cramping.'[172]

Moltmann asserts that the new creation will serve as the cosmic temple. The presence of God which dwells in this cosmic temple is the indwelling of his unmediated and direct glory. This indwelling presence makes heaven and earth new, and is also the really new thing in the new Jerusalem. Moltmann asserts, "God will 'dwell' among them. That is *the cosmic Shekinah*."[173] Thus God's unmediated and immediate presence interpenetrates everything. This eschatological indwelling of God in all things has two characteristics: holiness and glory. These are the goals of this creation. As Moltmann says, "The holiness and the glory of the eternal indwelling of God is the eschatological goal of creation as a whole and of all individual created beings. This gives eschatology a theological dimension and an aesthetic one."[174]

4.4 Critical Evaluation

Moltmann's eschatological, or panentheistic, vision of "God's being all in all" will be completed with the cosmic Shekinah. In other words, God's universal immanence will be completed. In this coming panentheistic state, God and the world will not be separable though they will still be mutually distinguishable. Through the communication of idioms God and the world will mutually participate in the attributes of the other. Although Moltmann is convincing that we can admit the participation of creatures in the divine attributes of eternity and omnipresence, to suggest the participation of God in limited time and the restricted space of creatures seems to me to seriously dilute God himself.

172 Moltmann, *The Coming of God*, p. 307f.
173 Moltmann, *The Coming of God*, p. 317.
174 Moltmann, *The Coming of God*, p. 318.

God, Space, and Spatiality-Case Studies

Polkinghorne opposes panentheism, because, to the extent that it enmeshes God and the world, panentheism threatens the mutually free relation both of God and his creation. Nevertheless, Polkinghorne likes Moltmann's idea of God's self-limitation toward creation. In order to defend panentheism, panentheists like Hartshorne often criticize classical theism by saying that "'If things were simply "outside" God, there would be a greater reality than God, God and the world.' Again, if one were to try to conceive of God as in some sense independent of the world, 'then God-and-what-is-other-than-God must be a total reality greater (more inclusive) than God.'"[175] Against this criticism, Polkinghorne maintains, "Since the world is only in being because of God's free making room for it, its existence in addition to God does not create a greater reality, for we are speaking ontologically and not arithmetically."[176]

Through the notion of God's initial self-limitation, Moltmann proposes a version of panentheism which seems more defensible to Polkinghorne:

> If creation *ad extra* takes place in the space freed by God himself, then in this case the reality outside God still remains *in* the God who yielded up that 'outwards' in himself. Without the difference between Creator and creature, creation cannot be conceived of at all: but this difference is embraced and comprehended by the greater truth which is what the creation narrative really comes down to, because it is the truth from which it springs: the truth that God is all in all.[177]

---

175 Colin E. Gunton, *Becoming and Being: The Doctrine of God in Charles Hartshorne and Karl Barth* (Oxford: Oxford University Press, 1978), p. 58.

176 Polkinghorne, *Science and Providence: God's Interaction with the World* (London: SPCK, 1989), p. 22. Unlike his fellow scientists as theologians Barbour and Peacocke, Polkinghorne has consistently opposed panentheism (*Scientists as Theologians*, p. 32). Even though he shares "a recognition of the need to correct classical theism's undue emphasis on the transcendent remoteness of God," Polkinghorne does not feel that this implies "a necessity to adopt panentheistic language." He states, "It simply requires a recovery of the balancing orthodox concept of divine immanence" (ibid., p. 33f). In contrary to these contentions, Polkinghorne seems to adopt two panentheistic ideas. One is the view of divine dipolarity between eternity and temporality (See *Scientists as Theologians*, p. 41, *Belief in God in an Age of Science*, p. 69, and *Science and Providence*, p. 80). The other is his view on creation as kenosis (See *Science and Providence*, p. 22 and *Science and Creation*, p. 61f).

177 Moltmann, *God in Creation*, p. 89.

120   *Transcendence and Spatiality of the Triune Creator*

This statement represents Moltmann's vision of eschatological panentheism. This is the final form which creation is to find in God. So the initial self-limitation of God assumes the accomplishment of the eschatological vision (i.e., "the glorifying, derestricted boundlessness in which the whole creation is transfigured").

Colin Gunton responds to "Moltmann's panentheistic notion of creation as kenosis" by presenting three reasons why it is unconvincing. First, according to Gunton, contra Moltmann, "it is possible to conceive a created world that is external to God and which does not yet exclude interrelationship and omnipresence." The second criticism that Gunton presents for the doctrine espoused by Moltmann is that there is little biblical support. Gunton writes, "there appears to be no reason why creation should involve a self-emptying of God if the universe is truly to be itself." Finally, Gunton opposes the metaphor of *kenosis*, because "it is a concept designed to deal with God's bearing in relation to a fallen world, not to be applied promiscuously to any of God's relations to the world."[178] The concept of Christian creation is the creation of the world outside of God. Even though Moltmann argues "the basic idea of this doctrine [divine *zimzum*] gives us the chance to think of *the world in*

---

178   Gunton, *The Triune Creator*, p. 142. *Kenosis* is the term to indicate the humiliation of Christ the Son. Rosemary Ruether applies the term to the Father. Even in Ruether, however, *kenosis* does not refer to the creation of the world. By the term, she suggests the possibility for God the Father to reveal himself in any other ways (Cf. Rosemary R. Ruether, *Sexism and God-talk: Toward a Feminist Theology* (Boston: Beacon Press, 1993), pp. 1–3). However, despite its lack of biblical support, many scholars hold this view of creation as kenosis. See J. Polkinghorne, ed., *The Work of Love: Creation as Kenosis* (Grand Rapids/Cambridge: Eerdmans/ SPCK, 2001). While rejecting "Arthur Peacocke's 'biological model', based on the metaphor of a maternal sustaining of the foetus in the womb, so that one has 'an analogy of God creating the world within herself'," Polkinghorne accepts Moltmann's account on creation as kenosis. He criticizes Peacocke's position as panentheistic idea. However, according to Polkinghorne, Moltmann's view draws the sting from Hartshorne's criticism of classical theism: "If things were simply 'outside' God, there would be a greater reality than God, God and the world" (Polkinghorne, *Science and Providence*, p. 22). But I think those two theologians' views are altogether panentheistic. The only difference is that, whereas Peacocke accepts the terms of maternal metaphor of feminist theologians such as Elizabeth Johnson, Moltmann does not adopt them [E. Johnson, *SHE WHO IS: The Mystery of God in Feminist Theological Discourse* (New York: Crossroad, 1995), p. 233f]. It seems to me that this is minor difference.

God, Space, and Spatiality-Case Studies

121

God without falling victims to pantheism,"[179] it leads him to panentheism which, according to Gunton, cannot finally be distinguished from pantheism.

Alongside Moltmann's idea of divine *zimzum*, two remaining issues should be addressed. One is the problem of God's embodiment, or his spatiality. Many modern theologians assert that God has his own temporality. Can we contend similarly for his spatiality? When the vision of God's being all in all is consummated, the mutual indwelling of God and the world is completed. Is such a completion possible for God to achieve even without having a body or spatiality? For Moltmann the spatiality of God is inevitable, although Moltmann does not contend boldly that God has a body. His contention that our suffering is in God,[180] his social Trinitarianism,[181] his adoption of the Jewish idea of *zimsum*, and his panentheistic vision of the mutual indwelling of God and the world are only possible, however, insofar as Moltmann has presupposed the spatiality of God. Furthermore, he explains the perichoretic concept of space as the model of mutual interpenetration of the three Persons. However, Moltmann does not agree with the Pannenberg's view of the Spirit as a field in order to explain the active presence of the transcendent God with his creation. Moltmann also does not refer to the world as the body of God, contra Grace Jantzen.[182]

Colin Gunton states that "the ghost of Newton still walks with greater or lesser influence in a number of modern trinitarian theologies, particularly in suggestions that the universe is to be understood as in some way (spatially) within the being of God."[183] After he deals with Moltmann's thought, Gunton introduces Robert Jenson's version that creation is within God: "Jenson is using the idea of roominess of God – a spatial image – to generate an alternative to Augustine's view of time."[184] Against these views, Gunton argues in the following way:

---

179 Moltmann, *The Trinity and the Kingdom*, p. 110.
180 Cf. Moltmann, *The Crucified God: The Cross of Christ as Foundation and Criticism of Christian Theology* (New York: Harper & Row, 1974).
181 Cf. Moltmann, *The Trinity and the Kingdom*.
182 Cf. Grace Jantzen, *God's World, God's Body* (Philadelphia: Westminster Press, 1984). Concerning criticism of this position, see Peacocke, *Theology for a Scientific Age*, p. 168; and Polkinghorne, *Science and Providence*, pp. 19ff.
183 Gunton, *The Triune Creator*, p. 140.
184 Gunton, *The Triune Creator*, p. 141.

122   *Transcendence and Spatiality of the Triune Creator*

> But if the world is truly to be the world, it needs to be 'outside' of God, not in such a relationship according to which it is in some way enclosed within God. For this reason, panentheism cannot finally be distinguished from pantheism, because it does not allow the other space to be itself.[185]

The second remaining issue is the relationship between the two terms of Moltmann's eschatological panentheism: "God in the world" and "the world in God." He describes the relationship between God and the world as "God *in* the world and the world *in* God."[186] Concerning this issue, a British interpreter of Moltmann, Richard Bauckham argues that, while "the world in God" represents his transcendence beyond it and its openness to his transcendence, "God in the world" expresses his immanence in it.[187] A Catholic feminist theologian, Elizabeth Johnson, also contends that "God in the world" indicates his immanence in it. She asserts that "All Christian speech about God, including classical theism with its heavy stress on transcendence, affirms that God dwells intimately at the heart of the world." However, concerning "whether the indwelling is reciprocal, that is, whether the world is likewise present in God," unlike Bauckham, she argues that "the spectrum of theological options includes at least two ways of saying no to that question and one way of coherently thinking through a yes." Whereas classical theism and pantheism deny it, panentheism affirms it.[188]

In his article "The World in God or God in the World?" Moltmann asserts that the great eschatological question is *"What is the ultimate goal? Is it the world in God, or God in the world?"* According to him, "God in the world" is the ultimate goal of his own theological thought whereas "the world in God" is ultimate for the theological perspective like that of Hans Ur von Balthasar. He contrasts his own view that "the eternal Trinity glorifies itself in the redemption of history and the consummation of the world" with von Balthasar's position that "the world's eschatology takes place in the eternal Trinity." Nevertheless, Moltmann maintains that there is no entirely unbridgeable contradiction between the two views. Thus he concludes as follows:

---

185 Gunton, *The Triune Creator*, p. 142.
186 Moltmann, *God in Creation*, p. 17.
187 Bauckham, *The Theology of Jürgen Moltmann*, p. 243.
188 Elizabeth A. Johnson, *SHE WHO IS: The Mystery of God in Feminist Theological Discourse* (New York: Crossroad, 1992), p. 230.

God, Space, and Spatiality-Case Studies

> The two can be related to one another like mirror images, as it were [...] the world will find space in God in a worldly way when God indwells the world in a divine way. That is a reciprocal perichoresis of the kind already experienced here in love: the person who abides in God and God in him (1 John 4:16). According to Paul, this presence of mutual indwellings is here called love, but then it will be called glory.[189]

Through his panentheism Moltmann intends to retain an emphasis on both divine transcendence and immanence. Like the other panentheists, he contends, "The one-sided stress on God's transcendence in relation to the world led to deism, as with Newton. The one-sided stress on God's immanence in the world led to pantheism, as with Spinoza."[190] As an alternative to these extremes, Moltmann proposes a trinitarian concept of creation and a panentheistic view: "The trinitarian concept of creation integrates the elements of truth in monotheism and pantheism. In the panentheistic view, God, having created the world, also dwells in it, and conversely the world which he created exists in him."[191] But Moltmann's primary concern is God's immanence in the world: God's transcendence, or the distinction between God and the world, must not be surrendered; "but an ecological doctrine of creation today must perceive and teach God's *immanence* in the world."[192] Thus Moltmann writes, "It is only the symbol of the world as work and as machine that makes God so transcendent that the immanence cannot be an equal counterpart for him, because it no longer partakes of his nature. The monotheism of the transcendent God and the mechanization of the world put an end to all ideas about God's immanence."[193] Finally, Moltmann discards "a dialectical structure of transcendence and immanence." Moltmann contends for "an immanent transcendence" and "a transcendent immanence."[194]

---

189 Moltmann, "The World in God or God in the World?" p. 41.
190 Moltmann, *God in Creation*, p. 98. This is a typical misunderstanding of panentheists including process theologians such as Charles Hartshorne.
191 Moltmann, *God in Creation*, p. 98.
192 Moltmann, *God in Creation*, p. 14.
193 Moltmann, *God in Creation*, p. 318.
194 Moltmann, *God in Creation*, p. 318. See also Steven Bouma-Prediger, *The Greening of Theology*, p. 117. I am indebted to him for my argument here.

124 *Transcendence and Spatiality of the Triune Creator*

5. Conclusion

This chapter has examined the thoughts of three theologians who have broached the topic of divine spatiality: Torrance, Pannenberg, and Moltmann. In comparison to the latter two, Torrance does not overtly discuss God's spatiality, but his discussion of temporality is highly suggestive of grist for our discussion. In comparison to the first, Pannenberg and Moltmann each assert some version of divine spatiality as well as temporality. And furthermore, for them, God's spatiality does not contradict his transcendence. Both of them try to explain God's intimate involvement within the world of space and time. Yet, their views on the Spirit as a field (Pannenberg) and divine *zimzum* as creation (Moltmann) betray a tendency toward panentheism. Even though they contend that panentheism holds together both the transcendence and the immanence of God in relation to the world, panentheists do fail to ensure divine transcendence. Therefore, although we can acknowledge that they have made a considerable contribution to expounding divine immanence within the universe, we must also see that they fail to maintain a creative tension or balance between God's transcendence and immanence.

Torrance has presented valuable insights for the discussion of divine spatiality. He has proposed the idea of the multi-leveled structure of human knowledge. By means of his thought on the redemption of space, he has sought to differentiate various kinds of space. He insists on the need to speak of divine temporality. Yet, he has not fully developed these thoughts. Along with Torrance, Pannenberg applies a field concept of modern physics in his doctrine of creation. Using terms from field theory, he attempts to describe the creative presence and activity of the Spirit in the world. He links the dynamic of the Spirit, as a field, to time and space. He asserts that the idea of God's transcendence rather demands space. By citing the omnipresence of God, Pannenberg holds that God must be conceived of as present in each part of space. This explanation of God's omnipresence leads him to a sacramental view of nature, and perhaps too far towards a veneration of nature. His theological system thus takes on a panentheistic hue. Moltmann, more overtly, calls himself a proponent for panentheism. He takes a traditional view on creation as a free act of God. But, he suggests that God, in some sense, needs the world. Employing the cabalistic term *zimzum*, Moltmann main-

God, Space, and Spatiality-Case Studies 125

tains that God makes room for creatures within himself before creation. Since creation, the world is in God. Using the term "cosmic Shekinah," he claims God eventually will be all in all. His eschatological vision expresses the ultimate fulfillment of "God in the world."

Despite some suggestive insights, Torrance's position on God's spatiality is too incomplete to determine his view of God's spatiality. By contrast, Pannenberg and Moltmann definitely argue for divine spatiality. But they seem to go too far. They accept the position of panentheism, whether overtly or implicitly. Between Torrance's tacitness and Pannenberg's and Moltmann's panentheistic suggestion, I think, there is room for a more valid presentation of God's spatiality. Thus the next chapter will suggest and examine various options and possibilities for speaking divine spatiality.

Chapter V

Divine Spatiality

1. Introduction

After surveying the concepts of space historically, I suggested the possibility of speaking of a divine spatiality that is also compatible with divine transcendence. My contention was that transcendence does not necessarily exclude spatiality. In the previous chapter, I examined three modern theologians' thoughts on divine spatiality in relation to divine transcendence. I noted that Torrance provides grist for this discussion by arguing for God's temporality, while being reluctant to admit God's spatiality. Yet, because his main concern is God's transcendence, he only briefly refers to the term "perichoresis" as the place of God. Moltmann, taking the perichoretic concept of space further, has proposed a kind of divine space in which we creatures can participate. Employing the term of divine *zimzum*, Moltmann contends for a space within God himself: before creation, God made a space for creation within him. Yet, Moltmann can be critiqued as going too far and thus sacrificing a firm statement of transcendence. Pannenberg has broached the subject of divine spatiality by attempting to explain the presence of the Holy Spirit by means of field theory developed by contemporary physics. Since, like Moltmann, his primary concern is also to emphasize divine presence, Pannenberg does not balance his view with a strong statement of divine transcendence. Both of them share the panentheistic position. Thus, my final verdict is as follows: On the one hand, while Torrance does not go far enough to develop a divine spatiality more positively, he succeeds in confirming the divine transcendence. On the other hand, while Pannenberg and Moltmann go further in the direction of divine spatiality, they do not succeed in allowing for God's transcendence.

This chapter will examine some positive options for God's spatiality. Some biblical passages refer to God as our dwelling place. It seems we

128   *Transcendence and Spatiality of the Triune Creator*

can adopt, at least figuratively, the option that God is space for his creatures. Another option is to say that God has space. God can use space as the vehicle of his presence in the world. I will support the position that God has his own space, and that the divine space is other than that of creatures. Because God is the creator of the world of space and time, he cannot strictly be tied to the structure of space and time. In this sense, God is not in the space and time of this world. However, God has his own space or his own spatiality. In order to have other space than that of creatures, does God have a body? To address this question, I will consider the option that the world is God's body. But, I will conclude that God is not a bodily being, nor does God need a body to interact with this world of space and time. Nevertheless God uses material beings, including the human body, as vehicles to express his glory. In order to use spatial beings, can God as a non-bodily being be in space? Even though God is not tied to space, I will contend, God can be present in space and act within it. By means of the terms "perichoresis" and "dimensional difference," I will argue that, even though he is transcendent over the world of space and time, God can be immanent, or present, within it. To be present in the world of space and time, the triune God in an important sense has his own spatiality.

Moltmann boldly maintains that God is space itself for us creatures. Moltmann identifies the perichoretic space of God with the triune God himself. Psalms 90.1 is one of the biblical passages that refers to God as a place, or space. Yet, I argue this verse is not to be interpreted literally. Thus, I will hold that we should speak figuratively when we say, God is space (section 2). Newton's well-known contention that his absolute space is the *sensorium Dei* is generally regarded to mean that God has space as a sense organ. This is a panentheistic idea and thus should be discarded. But if we do not take Newton's *sensorium Dei* literally, we can contend that God uses space as the means of his presence in the world of space and time. One more option is possible. That is, we may say that God has his own space, or God is in some sense spatial. I will support this latter view in section 3. Since this view is necessarily related to the corporeality of God, I will deal with the problem of the body in Christian tradition. To say "God is spirit" is not to imply that God is a non-bodily being. Rather it means that God is the relational and active God. I will also refer to both the bodily resurrection as our final hope, and the humanity of the ascended Jesus as viewed by the Reformed tra-

Divine Spatiality

dition. Regardless of the validity or invalidity of McFague's contention of the world as the body of God, Christians take our bodily life in the world seriously (section 4). In this sense, Dyrness's book, *The Earth is God's*, is valuable. He presents the three categories of relationship, agency, and embodiment. All of them are grounded in the presence of the triune God and provide windows through which to illuminate our corporeal life in the world (section 5). Tracy suggests that it is possible to speak of God's action in the world without admitting that he has a body. Yet, it is important to note that Tracy's main target is process theologians' contention that the world is the body of God, not Dyrness's third category of embodiment (section 6). By means of the terms, "perichoresis" and Schwarz's dimensional model, I will argue that God is present in space of this world in a way that does not compromise his transcendence over the world (section 7).

2. God Is Space

Moltmann's concept of perichoretic space provides Christian theology today with some valuable suggestion of how God's spatiality is different from that of creatures, despite the problems with his idea of divine *zimzum*, which implies that creation takes place within the space made available by God's self-contraction. But this panentheistic idea does not allow sufficiently divine transcendence.

In the last chapter of the final volume of his Contributions to Systematic Theology, *Experiences in Theology*, Moltmann indicates the change of his focus from the early theology of time to the later theology of space:

> Ever since my work on an ecological doctrine of creation and a social doctrine of the Trinity, I have tried to expand my theological world, which had been one-sided in its orientation towards time. I now tried to extend it through the concepts of space and 'home,' the Shekinah and the perichoresis, reciprocal indwelling and the coming to rest in one another. The link between my early theology of time and this later theology of space was for me the discovery of the fundamental importance of

130 *Transcendence and Spatiality of the Triune Creator*

the sabbath for the doctrine of creation and for the messianic expectation of the future.[1]

He titles the third section of the chapter as "From the historical hope in God's promise to the spatial experience of God's indwelling."[2] By this phrase Moltmann presents his own summary of the development of his theological thought from *Theology of Hope* up to the present. The concept of perichoresis, of course, plays an important role in the argument of that chapter. Here is Moltmann's explanation of perichoresis:

> The idea of mutual indwelling, *perichoresis*, goes back to the theology of the Greek Fathers, and makes it possible to conceive of a community without uniformity and a personhood without individualism [...] In *christology*, perichoresis describes the mutual interpenetration of two different natures, the divine and the human, in Christ, the God-human being [...] In *the doctrine of the Trinity*, perichoresis means the mutual indwelling of the homogeneous divine Persons, Father, Son and Spirit [...] Jesus and God the Father are not *one and the same*, they are *at one* – a unity – in their mutual indwelling.[3]

This explanation is not a new one. Moltmann has already dealt with the concept of perichoresis in his previous works. In this sense, the above cited explanation of perichoresis is merely a repetition of his previously stated thought. But Moltmann then deploys a further developed insight on the concept of perichoresis. He avers as follows: "every Trinitarian Person is not merely *Person* but also *living space* for the two others [...] Consequently we should not talk only about the *three trinitarian Persons*, but must at the same time speak of the *three trinitarian spaces* in which they mutually exist."[4]

Moltmann's eschatological vision is described as "God's being all in all" (1 Cor. 15:28). In this state, a mutual indwelling of God and human beings will be realized. This is the meaning of 1 John 4:16, as Moltmann interprets the verse: "He who abides in love abides in God and God in him." Moltmann refers to the deification of all created beings in the eternal presence of the triune God. So far as this concerns his eschatological thought, it is nothing new. But, at this point he contends, "all created

1   J. Moltmann, *Experiences in Theology: Ways and Forms of Christian Theology* (Minneapolis: Fortress Press, 2000), p. 314.
2   Moltmann, *Experiences in Theology*, p. 313.
3   Moltmann, *Experiences in Theology*, p. 316.
4   Moltmann, *Experiences in Theology*, p. 318f.

Divine Spatiality 131

beings will find their 'broad place where there is no more cramping' (Job 36.16) in the opened eternal life of God, while in the glorified new creation the triune God will come to his eternal dwelling and rest, and to his bliss."[5] Thus finally we will be in the perichoretic unity of the Trinity which lies in the perichoretic co-workings of the three divine Persons. The trinitarian unity of God is open to the world. The church ultimately exists in the tri-unity of God. Moltmann boldly insists that the triune God is experienced as the broad place, or the space, indwelt by us:

> The open space of the perichoretic community of the triune God is the divine living space of the church. In the community of Christ and in the energies of the life-giving Spirit we experience God as the broad place which surrounds us from every side and brings us to the free unfolding of new life. In the love which affirms life we exist in God and God in us. The church is not just the space for the indwelling of the Holy Spirit. It is the space indwelt by the whole Trinity. The whole Trinity is the living space of the church, not just the Holy Spirit.[6]

Here if we take these sayings of Moltmann literally, we are confronted with at least one error. Moltmann identifies the perichoretic space of God as the triune God himself. Moreover, he says we can exist in the perichoretic space of the triune God. That is the meaning of the second dimension in Jesus' prayer, "that they may also be in us" (Jn. 17:21). I think, however, we creatures do not exist in the triune God in the literal sense.

Moltmann describes the new life which we enjoy in the triune God as follows: "We are drawn into the Trinitarian history of God with the world, its creation, its redemption and its glorification, through baptism in the name of the triune God [...] The person who confesses the triune God begins to live 'in him.' We experience ourselves in God and God in us (I John 4.16). That is the new, true life."[7] Yet, unlike Moltmann, I contend that we cannot exist in the triune God, but we can dwell in the perichoretic space of God. It is, in other words, by the Spirit that we can exist in the triune God.

In the scriptures, we find some passages that refer to God as a place or even as space. One of them is Ps. 90.1: "You have been our dwelling place throughout all generations." It should suffice to examine Calvin's

5   Moltmann, *Experiences in Theology*, p. 323.
6   Moltmann, *Experiences in Theology*, p. 330.
7   Moltmann, *Experiences in Theology*, p. 312.

132   *Transcendence and Spatiality of the Triune Creator*

explanation of this verse. He notes, it is clear that the condition of all human beings is unstable upon earth. Abraham and his posterity were sojourners and exiles, wandering in the land of Canaan living only by sufferance from day to day after they were brought into Egypt. Thus Calvin states, "it was necessary for them to seek for themselves a dwelling-place under the shadow of God, without which they could hardly be accounted inhabitants of the world."[8] The Lord gave them grace and shielded them with his hand as they were constantly in peril. In extolling this grace, Moses represents God as an abode, or indwelling-place for these poor nomads and fugitives. Thus Calvin understands this verse as one that expresses the grace of the Lord by which the Israelites dwelt under the wings of his protection, rather than as a philosophical proposition that God is a kind of place or space. The emphasis in this verse is on God's protection, or God's action on his people's behalf. In a sense then, God's "place" here is a metaphor for God's grace and presence. As Scripture says elsewhere: "The eternal God is your refuge, and underneath are the everlasting arms" (Deut. 33:27), God is metaphorically praised as a space, or place, for us. Similarly, Paul told the Athenians, "in him we live and move and have our being" (Acts 17:28). Though the option that God is space in the literal meaning should be discarded, the metaphorical use of it is thoroughly possible.

3.   God Has Space

To make a statement for the position that God has space there are at least two options. One is to say that God has the world of space-time continuum as something like his sense organ. The other is to say that God has his own space, different from that of creation, in the same way that he has his own time. For Newton, space and time are *sensorium Dei* with which God perceives the universe. God is the one who constitutes time and space by existing always and everywhere. For Newton, as Gunton states, "space and time are the 'places' where God is omnipresent to the

---

8   J. Calvin, *The Comprehensive John Calvin Collection, Commentary on the Psalms*, vol. 2, p. 150f.

Divine Spatiality 133

world: the focuses of mediation where God at once creates and experiences the world."[9]

Pannenberg argues that Newton did not intend to ascribe to God an organ of perception, as his well-known conception of space as sensory of God seems to imply God does not need this, because he is omnipresent. According to Pannenberg, Newton's idea was easily mistaken as indicating some monstrously pantheistic conception of God similar to that found in Leibniz's polemics against Newton.[10] Newton described space as the *sensorium Dei* in his *Opticks* in 1706. Leibniz criticized Newton as a half-materialist. For Leibniz, Newton came close to describing God as a corporeal being. Leibniz further attacked Newton's conception of God: "Newton demotes the perfection of the divine work of creation as well as its creator by presenting the world as a machine constantly in need of repairs."[11]

In his discussion with Leibniz, Samuel Clark, the theological advocate of Newton, designated Newton's absolute space as a property of God. Leibniz objected to this point by noting that things may change their place in space so that space cannot be their property. However, Pannenberg contends, Leibniz's critique of Clarke's conception of space as property betrays Leibniz's misunderstanding of Clark. Behind the misunderstanding lie conflicting notions of space: "Leibniz thought of space with Descartes as a filled space and not as an empty space [...] Newton and Clarke, however, thought of absolute space as empty, and this is decisive for their conception of the relationship of God and space."[12] Citing the saying of Alexander Koyré, that "for Newton, it is just this presence that explains how God is able to move bodies in space by his will – just as we move our body by the command of our will," Pannenberg proposes that Newton's contention of space as *sensorium Dei* be read as an implicit rejection of Descartes's conception of God which, according to Newton, separated God from the world and made God absent from the world. For Newton, Pannenberg avers, "God is

---

9   Gunton, *The Triune Creator*, p. 128.
10   Pannenberg, *Toward a Theology of Nature*, p. 42. In this case, I think the term "panentheistic" is better than "pantheistic."
11   Pannenberg, *Toward a Theology of Nature*, p. 59.
12   Pannenberg, *Toward a Theology of Nature*, p. 61.

134   *Transcendence and Spatiality of the Triune Creator*

present for the creatures through space."[13] In this vein, Pannenberg interprets Faraday's field theory, as well as Einstein's relativity theory, as an underlying renewal of the deeper intentions of Newton himself.[14]

Pannenberg's claim that Newton did not understand the *sensorium Dei* as God's organ of perception but rather as the medium of his creative presence at the finite place of his creatures excludes the option that God has space as his sense organ. Pannenberg takes figuratively Newton's contention of absolute space as the *sensorium Dei*. For Pannenberg, rather, space is the medium of God's presence. In other words, God uses space to be present in the world. This understanding of Pannenberg on space is valuable. But in order to defend it, I would suggest, Pannenberg needs to discard Newton's conception of absolute space and adopt a relational concept of space. Since the concept of absolute space is intimately related to the receptacle idea of space, the infinite God cannot be present in the finite place of his creatures.

Before we go to the second option, we should heed Polkinghorne's caution against regarding quantum fields as the "sensorium of God": "We can think of the field as the letting-be that enables the is-ness of its particle excitations to occur [...] To take it more seriously than that would be to regard quantum fields as the 'sensorium of God' (in the way that Newton was emboldened to talk about absolute space), a panentheistic idea which does not commend itself to me. When all is said and done, quantum fields are simply creatures."[15] Contrary to Pannenberg, Polkinghorne takes literally the *sensorium Dei* of Newton. But if we give the *sensorium Dei* a literal interpretation, it becomes a panentheistic statement. Though we can discard the option that God has space as something like a sense organ, we can accept the view that, as Pannenberg puts it, God uses space as the medium of his creative presence.

---

13   Pannenberg, *Toward a Theology of Nature*, p. 62. So we can assert that even Newton's notion of the world as a machine, though repugnant to us, was an advance on what preceded him.

14   Pannenberg, *Toward a Theology of Nature*, p. 64. At the same time, however, Pannenberg indicates Leibniz's inspiration on Faraday "in seeking the unity of the nonbodily force which determines the natural processes but escapes mechanical description" (ibid.).

15   Polkinghorne, *Science and Creation: The Search for Understanding* (London, SPCK, 1988), p. 58.

Divine Spatiality 135

While he acknowledges that God is supreme over all space and time, Donald Bloesch contends that God has his own space and time. God's spatiality is not contradictory but identical with his being. Thus Bloesch avers that God has his own spatiality. Because he has time and space within himself, God does not contradict himself by creating historical time and creaturely space. Bloesch asserts that God's eternity includes the potentiality of time and space. He states, "Our time and space reflect the absolute time and space that constitute the very being of God and that alone are enduring."[16] This contention regarding God's space is closely related to that of God's time.

Traditionally God's eternity has meant his being outside of time. Since he is eternal, God is a timeless being. But with this understanding, it is difficult to find meaning in the biblical records on God's activity within time. Since Barth, therefore, many theologians have developed views on divine temporality. Reflecting on these developments, and asking whether we can say that God has a time which moves at a different speed from ours, Gerald Bray asserts, these temporalist views create just as many problems for a doctrine of divine intervention in the world as we find in the doctrine of a timeless eternity.[17] Paul Helm also strongly opposes the temporalist views and holds to a description of an eternal God in which God is without time.[18] One of Helm's main arguments against the divine temporality is that, if we accept that God is a temporal being, we should admit that God is also a spatial being: "one consequence of God being in time is that God is finite because there is a relevantly parallel argument for God being in space which carries this consequence."[19] For Helm, it seems easier to refute divine spatiality than divine temporality. As he says, "One reason for thinking that the idea of God being in space is unacceptable is that such a god would be finite, or

16   D. Bloesch, *God the Almighty* (Downers Grove, IL: IVP, 1995), p. 88. On this point, Bloesch seems to make a mistake similar to that of Moltmann. They both misunderstand absolute space as a kind of divine space. See Moltmann, *God in Creation*, p. 156f. Moltmann speaks of, "first, the essential omnipresence of God, or absolute space; second, the space of creation in the world-presence of God conceded to it; and third, relative places, relationships and movements in the created world."

17   G. Bray, *The Doctrine of God*, p. 84.

18   P. Helm, *Eternal God: A Study of God without Time* (Oxford: Clarendon Press, 1988).

19   Helm, *Eternal God*, p. 42.

136   *Transcendence and Spatiality of the Triune Creator*

at least the argument used to show that he is in space would show that he is in space finitely or boundedly, and so is finite."[20]

Peter Kreeft and Ronald Tacelli seem to share the view of Helm. In recent years, they argue, some Christian philosophers and theologians have dismissed the doctrine of God's eternity. According to Kreeft and Tacelli, there are three options: One is that God exists everlastingly in time. The other is that, since the creation of the world, God exists in time only. The third option is panentheism. Kreeft and Tacelli describe panentheism as one way of making God temporal. However, they acknowledge that there are some orthodox theologians who maintain that God is somehow in time, while avoiding panentheism.[21]

In his article, "God's Place in a Space Age," Hans Schwarz proposes the metaphor of a "dimensional relationship" between God and the world: "If we now assume that God is related to us in a dimensional way, being present in a way in which he is dimensionally higher than we are, God would embrace all our available possibilities in space and time plus possibilities which are not available to us in our present dimension. Thus both elements, God's presence in our space-time continuum and God's superiority to it, could be maintained."[22] In other words, this model makes sense both divine transcendence and immanence alike: "God would be present in our dimension (i.e., four-dimensional space-time continuum) without being contained by it and he would transcend it without being absent."[23] Schwarz refers to Torrance who seems to hold this view. Schwarz quotes the following passage from Torrance's *Space, Time and Resurrection*:

> In our investigation of nature we frequently come across a set of circumstances or events which do not seem to make sense for we are unable to bring them into any coherent relations with one another, but then our understanding of them is radically altered when we consider them from a different level, for from that point of view they are discerned to form a distinct, intelligible pattern. This can happen when an additional factor is included at the original level which helps us to solve the puzzle, but often the all-important additional factor must be introduced from a higher level,

---

20   Helm, *Eternal God*, p. 42.
21   Kreeft and Tacelli, *Handbook of Christian Apologetics*, p. 94.
22   Hans Schwarz, "God's Place in a Space Age," *Zygon* 21/3 (September 1986), p. 365f.
23   Schwarz, "God's Place in a Space Age," p. 367.

Divine Spatiality 137

which means that the coherent pattern of the circumstances or events we are study-
ing is reached only through a dimension of depth involving cross-level reference.[24]

Although this dimensional relational schema is not set forth as a panacea
in relating God and the world, Schwarz contends that "it does suggest
one model in which God's place in relation to the world can be recon-
ceptualized in a new and intelligible way."[25]

If rightly understood and interpreted, Moltmann's perichoretic concept of space can be thought of as an example of the dimensional relational model which Schwarz proposes. It is not problematic to speak of
God as having perichoretic space, as his own space different from that of
us creatures. But it is still not clear what is meant by the contention that
God has his own space, God is spatial, or God has spatiality. At the creaturely level, we do need to have bodily existence to have or occupy
space. For, on the creaturely level, to have space means to occupy a
place. What about God?

4. The Body of God

The Johannine statements that "God is Spirit" (John 4.24) and that "God
is love" (1 John 4:8) are two of the few biblical verses that explicitly
characterize the divine essence.[26] Kreeft and Tacelli present a traditional
summary of what it means to say God is spiritual:

> God is not a material being. To be a material being is to be a body of some kind.
> But a body is always limited and subject to change. To be subject to change in this
> way is *not to be* what one *will become*. And therefore to be subject to change in-
> volves *nonbeing*. And since to be a body is to be subject to change, therefore to be
> a body involves nonbeing. Now God is the limitless fullness of being, so God can-

---

24 Torrance, *Space, Time and Resurrection*, p. 188.
25 Schwarz, "God's Place in a Space Age," p. 367.
26 Cf. Pannenberg, *Systematic Theology*, vol. 1, trans. G. W. Bromiley (Grand Rapids:
Eerdmans, 1991), pp. 294, 395f.

138 *Transcendence and Spatiality of the Triune Creator*

> not be a body. In fact, God cannot be material at all – at least not as matter is normally understood. God must be immaterial, that is, spiritual.[27]

This understanding of God's spirituality is intimately related to the divine simplicity of God. God is one and indivisible. He does not encompass within himself disparate parts or quantities. And He cannot be analysed or subdivided into parts. Thomas V. Morris defines the doctrine of divine simplicity as "the claim that, in his innermost being, God must be without any sort of metaphysical complexity whatsoever."[28] In Morris's view, this doctrine implies a threefold denial:

> (1) God is without any spatial parts (the thesis of spatial simplicity), (2) God is without any temporal parts (the thesis of temporal simplicity), and (3) God is without the sort of metaphysical complexity which would be involved in his exemplifying numerous different properties ontologically distinct from himself (the thesis of property simplicity).[29]

Morris contends that, even though the thesis of temporal simplicity is very controversial, the thesis of spatial simplicity is endorsed by the vast majority of traditional Christian theists, because of considerations deriving from perfect-being theology. The idea behind this theology is that "God is not a physical object, so he does not have physical, or spatially located, parts."[30]

Gerald Bray, an Anglican theologian, explains that "the term 'simplicity' has sometimes been used in a wider sense, to argue (quite correctly) that God is 'without body, parts or passions,' as Article I of the Church of England puts it."[31] He then further demonstrates that this understanding is an extension of its original meaning and should not be confused with it. Bray's point provides grounds for criticizing Morris, for Morris indeed confuses the doctrine of divine simplicity with the thesis of property simplicity.[32] Bray proposes God's simplicity be under-

---

27 Kreeft and Tacelli, *Handbook of Christian Apologetics*, p. 92.
28 Thomas V. Morris, *Our Idea of God: An Introduction to Philosophical Theology* (Dowers Grove, IL: IVP, 1991), p. 113.
29 Morris, *Our Idea of God*, p. 114.
30 Morris, *Our Idea of God*, p. 114.
31 Bray, *The Doctrine of God*, p. 94f.
32 Concerning the thesis of property simplicity, Morris, *Our Idea of God*, p. 114, says, "if God were like us in exemplifying properties distinct from himself, then he would depend on those properties for what he is, in violation of divine aseity."

Divine Spatiality

stood only in relation to God himself: "Nor does simplicity mean that God is the primordial substance out of which everything else is made. Creation is not an extension or a corruption of God's being, but something else which is quite a different reality."[33] Donald Bloesch also believes that by divine simplicity God's relation to the world is not to be understood as an amalgamation or combination with the world process. He writes, "This does not imply that God is monochrome or solitary. He exists as a triune fellowship within himself and relates himself to the world in a multitude of ways. He is not simply personal but interpersonal."[34] Bloesch argues that God is not a material being, but God can assume a material form: "He [God] has done so in the incarnation of his Son, Jesus Christ. His Son was resurrected not as a pure spirit but in a bodily form that carried the earthmarks of materiality. The resurrection means not liberation from the flesh but glorification of the flesh. Yet this glorification involves a transformation and elevation of flesh into a spiritualized form of existence."[35]

For Stanley Grenz, to say, "God is spirit," is not to imply that God is a non-bodily being but that God is the source of life and that he is a relational God. Grenz elaborates as follows:

> By declaring "God is spirit' we acknowledge that God is the source of all life. God is the one who bestows life on his creatures – on each living thing but most significantly on humans [...] Lying behind God's relationship to the world as the giver of life is a prior internal divine relationship, an eternal relationship within the triune God [...] To say "God is spirit," therefore, is to speak about the relational God [...] To say "God is spirit" is to acknowledge that the vitality of the triune God overflows to creation. The God who is dynamic activity within the eternal Trinitarian life relates to the world as the source and sustainer of created life.[36]

Even though "the vast majority of philosophically sophisticated theists throughout history have argued that God can be neither be identical with any physical object nor even be said in any sense to have a physical body,"[37] those arguments are not crucial.

---

33 Bray, *The Doctrine of God*, p. 94f.
34 Bloesch, *God the Almighty*, p. 90.
35 Bloesch, *God the Almighty*, p. 89.
36 Grenz, *Theology for the Community of God*, p. 83.
37 Morris, ed., *The Concept of God* (Oxford: Oxford University Press, 1987), p. 13.

140   *Transcendence and Spatiality of the Triune Creator*

In his essay, "God's Body," William J. Wainwright states, "The belief that God has a body is by no means uncommon in the history of religious thought, nor is that belief confined to the primitive and unsophisticated. The Manichaeans, Tertullian, certain Egyptian anchorites, and the Mormons have all maintained that God has (or is) some particular body."[38] Wainwright thus begins his essay by reviewing the standard arguments against God's being a body. He argues that it is only with the presupposition that classical theology is right, that we can discard the view that God is a body.

Then, Wainwright considers the weaker claim that God *has* a body, and reviews a series of classical arguments to the contrary. Traditionally, the possession of a body has been thought to be an imperfection: "If one cannot possess a body without being subject to these evils, then no perfect being possesses a body and, hence, God does not possess a body." Against this argument, Wainwright says, "Neither Adam's body nor the bodies of resurrected humanity will be subject to them."[39] Under the presupposition that some divine attributes of classical theism are right, Wainwright discards the option that God has a body.[40]

Wainwright goes on to point out that there is an ambiguity in the thesis that the world is God's body: "God's body may be identified with the world in its entirety or it may be identified with the physical or material aspect of the world."[41] Morris presents the following summary of Wainwright's arguments: "After trying out a number of models of embodiment which seem clearly incompatible with many of the commitments of classical theism, Wainwright makes the surprising suggestion that a peculiarly platonic conception of the body-soul relation in human embodiment can be used to model a divine embodiment in the world in such a

---

38   W. J. Wainwright, "God's Body," in Morris ed., *The Concept of God*, p. 77.
39   Wainwright, "God's Body," p. 78.
40   Wainwright, "God's Body," p. 80. Concerning the orthodox view that God does have a (particular) body, viz. the body of Jesus, Wainwright asserts the following: "The fact that the body of Jesus is extended, spatially circumscribed, contingent, destructible, etc., does not imply that God (a divine subject in its divine nature) is extended, spatially circumscribed, contingent, destructible, etc. In short, Christians believe that the Incarnation on the one hand, and God's immateriality and freedom from physical limitations on the other, are compatible. Whether this attempted resolution of an apparent incompatibility is successful or not is, of course, a moot point."
41   Wainwright, "God's Body," p. 80.

Divine Spatiality

way as to comport with most classical theistic concerns about the nature of deity."[42] Indicating that "Hartshorne's position involves the acceptance of a Whiteheadian analysis of the soul-body relationship and a rejection of much classical theology," Wainwright finally asserts that "those who find this too high a price to pay must either adopt the 'Platonic' (or some similar) model, or reject the view that the world is God's body."[43]

Though the conclusion of Wainright's article is not acceptable to me, his contention that God is positively related to body is suggestive. In Christianity, there is a deep-rooted hostility to the body. This tendency is not based on a balanced interpretation of the biblical data. It is, rather, a result of the influence of Hellenism upon patristic theology. Even though I oppose any option that God is body or God has body, I will argue that God is very concerned about our bodily life on the earth, because he is also the creator of the body, and he himself assumed a body in Jesus Christ.

Gerald Bray aptly holds that "it is not at all obvious that the God of the Bible is without a body, parts or passion, even if these are not the same as their human counterparts."[44] Some process theologians suggest that the universe itself is God's body, but it is very difficult to see what biblical evidence can be cited in support of this idea. A more common option is "the assertion that God has a spiritual body of some kind, which is not visible nor entirely analogous to the human body."[45] Donald

---

42 Morris, ed., *The Concept of God*, p. 13.
43 Wainwright, "God's Body," p. 87.
44 Bray, *The Doctrine of God*, p. 95.
45 Bray, *The Doctrine of God*, p. 96. Concerning angels' corporeality, Bray says the following: "An angel might seem to be 'incorporeal' to us, but it was not so to John, or to his mediaeval successors, who believed that spirits could and did occupy a space, since otherwise they would be co-extensive with God himself. This is one reason why the question of how many angels could stand on the head of a pin did not seem as silly to them as it may do to us" (ibid., 97). Commenting on the role of angels, T. F. Torrance also refers to this medieval discussion: "Angels themselves are not restricted or hemmed in by space and time as we are, but they direct us at the boundary of our earthly existence within space and time and at the frontiers of human knowledge to listen to the voice of God from beyond and to distinguish it from all earthly voices." Then in a footnote Torrance explains as follows: "This is the significance of the medieval discussion about how many angels can dance on the point of a pin! In their tangential relation to our world they are

142   *Transcendence and Spatiality of the Triune Creator*

Bloesch maintains that God has a spiritual body, just as he has his own space and time. Against the Hellenistic view, he argues that, in biblical parlance, spirit is not opposed to body. He defines spirituality in biblical perspective as the descent of the spiritual into the material. Thus spirituality involves meeting material and genuine human needs but with a spiritual purpose and goal. Bloesch avers that "the spirituality of God is not to be confused with a spirit world beyond the material; instead it is to be seen as the light of the presence of God that breaks into the darkness of our world and sets us free to serve in the very midst of the degradation that surrounds us." For Bloesch, "God is a spirit, but he includes the phenomena of nature, humanity and history within his comprehensive vision."[46]

Sallie McFague presents a more radical view of God's body. She holds that "the idea of God's embodiment [...] should not be seen as nonsense; it is less nonsense than the idea of a disembodied personal God."[47] She proposes the model of the universe as God's body. To her, this model is "a way of expressing both radical transcendence *and* immanence, but in a fashion that limits our perception and knowledge to the back of God." That is to say, "[t]he body of God is not the human body nor any other body; rather, all bodies are reflections of God, all bodies are the backside of divine glory."[48] In relation to the issue of God's embodiment, McFague explains pantheism, traditional theism, and panentheism as follows: "Pantheism says that God is embodied, necessarily and totally; traditional theism claims that God is disembodied, necessarily and totally; panentheism suggests that God is embodied but not necessarily or totally. Rather, God is sacramentally embodied: God is mediated, expressed, in and through embodiment, but not necessarily or totally."[49] For her, thus, the world (universe) is God's body and God is its spirit.

---

    like spaceless mathematical points" [Torrance, *The Christian Doctrine of God* (Edinburgh: T&T Clark, 1996), p. 230].

46   Bloesch, *God the Almighty*, p. 89.

47   Sallie McFague, *Models of God: Theology for an Ecological, Nuclear Age* (Philadelphia: Fortress Press, 1987), p. 71.

48   McFague, *The Body of God: An Ecological Theology* (Minneapolis: Fortress Press, 1993), p. 133f.

49   McFague, *The Body of God*, p. 149f. Elsewhere she presents the following definition of panentheism: "it is a view of the God-world relationship in which all things

Divine Spatiality 143

McFague is less radical than Grace Jantzen in that McFague indicates her own model of "the world as God's body" as a model to explain the relation between God and the world.[50] She contends that the world as God's body is a kind of metaphor. Quoting Exodus 33, "And you shall see my back; But my face shall not be seen," McFague states that, "This body is but the backside of God, not the face; it is the visible, mediated form of God, one that we are invited to contemplate for intimations of divine transcendence."[51] She claims that "Metaphor always has the character of 'is' and 'is not': an assertion is made but as a likely account rather than a definition."[52] In relation to metaphor, McFague asserts, a model is a metaphor with "staying power" or "sufficient stability and scope so as to present a pattern for relatively comprehensive and coherent explanation."[53] Thus, like other feminist theologians, she holds that, since God the Father is also a metaphor, we can freely change it to speak of God the Mother. McFague argues that, "since no language about God is adequate and all of it is improper, new metaphors are not necessarily less inadequate or improper than old ones."[54]

Colin Gunton criticizes McFague's metaphorical theology.[55] He states that, "it does not follow from saying that if God is only metaphorically – and so both is and is not – Father, he is also justifiably said to be mother."[56] For Gunton, unlike McFague, metaphors are not odd, unusual,

---

have their origins in God and nothing exists outside God, though this does not mean that God is reduced to these things" (ibid., 72).

50 Cf. Grace Jantzen, *God's World, God's Body* (Philadelphia: Westminster Press, 1984). Arthur Peacocke differentiates McFague's position from Jantzen's view in this way: While the former does not lose sight of the metaphorical character of the world as God's body, the latter regards such language in a much stronger ontological sense [A. Peacocke, *Theology for a Scientific Age: Being and Becoming – Natural, Divine, and Human* (Minneapolis: Fortress Press, 1993), p. 168].

51 McFague, *The Body of God*, p. 156.

52 McFague, *Models of God*, p. 33.

53 McFague, *Models of God*, p. 34.

54 McFague, *Models of God*, p. 35.

55 See C. E. Gunton, "Proteus and Procrustes: A Study in the Dialectic of Language in Disagreement with Sallie McFague," in Alvin F. Kimel, Jr. ed., *Speaking the Christian God: The Holy Trinity and the Challenge of Feminism* (Grand Rapids: Eerdmans, 1992), pp. 65–80; and C. E. Gunton, *The Actuality of Atonement: A Study of Metaphor, Rationality and the Christian Tradition* (Grand Rapids: Eerdmans, 1989), 40–47.

56 Gunton, "Proteus and Procrustes," p. 66.

144 *Transcendence and Spatiality of the Triune Creator*

improper or merely decorative but a pervasive part of our experience. He states, metaphors are "a, if not the, clue to what language is and does."[57] Contrary to the contention of feminist theologians, including McFague, that we can replace God the Father with God the Mother, Gunton opposes to them. It is not theologically appropriate to call God the Mother. He claims, "The fatherhood of God has nothing to do with maleness but has to do with patterns of relationality revealed and realized in Jesus."[58]

In his "Synodical Report of Committee to Study Inclusive Language for God" for the Christian Reformed Church in North America, John W. Cooper presents three general solutions to the theological problem of language for God: 1. Language for God is analogical and can truly describe. 2. Language for God is figurative and can truly describe. 3. Language for God is figurative and cannot truly describe.[59] The first two are acceptable but the last one is not. Cooper cites McFague as the most articulate champion of the last position. Concerning her view, he maintains that, "Since all our language about God is metaphorical, she argues, whatever we say about God amounts to no more than claiming 'it is *as if* God is like this' (*Models of God*, 70). We cannot say what God is really like."[60] Even though this view recognizes God's transcendence, it is in fact a kind of religious and theological agnosticism.

Colin Gunton argues that an urgent need of our times is for an ethic of creation, especially ecological ethics. He assesses this problem as follows: "If the universe is treated simply as a mechanism, within the control of human reason and without direct reference to God, those dimensions of it which are not mechanical, and particularly its living creatures, are treated with despite, and react in such a way that our own existence is threatened."[61] Following the view of Pannenberg, Gunton asserts that the ecological crisis is a result of the modern abandonment of a religious view of the world, not a result of the Christian teaching about hu-

---

57 Gunton, *The Actuality of Atonement*, p. 32.
58 Gunton, "Proteus and Procrustes," p. 78.
59 John W. Cooper, "Synodical Report of Committee to Study Inclusive Language for God" (unpublished report to the Synod), p. 34f.
60 Cooper, "Synodical Report," p. 35. Concerning his more detailed view on inclusive language for God, see Cooper, *Our Father in Heaven: Christian Faith and Inclusive Language for God* (Grand Rapids: Baker Books, 1998).
61 C. E. Gunton, "The Doctrine of Creation," in C. E. Gunton ed., *The Cambridge Companion to Christian Doctrine* (Cambridge University Press, 1997), p. 154f.

Divine Spatiality

man "dominion" over creation. According to Gunton, the so-called "ecofeminist" theologians including Sallie McFague, have lurched in the opposite reactive direction to this crisis. These positions, however, tend toward a view of "the divinity of the universe," according to Gunton. More specifically, he presents the following criticism: "These doctrines tend to assimilate the doctrine of redemption to the doctrine of creation."[62] This assimilation is one way in which panentheism slides into pantheism.[63] Thus, Gunton says, "A pantheism in which we are simply a function of the world's process is a world in which our otherness, our capacity to be ourselves, would be taken away. Rather than reacting in this way, we need to move from seeing creation as a mere given to receiving it as gift to be cherished, perfected and returned: as grace evoking gratitude."[64]

Gunton's criticism of McFague is well founded. But there are some sound points among McFague's contentions. Citing Calvin's claim that "nature is the stage for salvation history," McFague holds that "[i]n this way of viewing the relation between creation and redemption, creation plays no critical role."[65] If her interpretation of Calvin's view is correct, it seems to me that his view about creation is not proper. So despite McFague's error in asserting "the central importance of creation," she is correct in highlighting that

> creation – meaning our everyday world of people and cities, farms and mountains, birds and oceans, sun and sky – is the place where it all happens and to whom it happens. Creation [...] is God's place and our place, the one and only place. Creation is not one thing and salvation something else; rather, they are related as scope and shape, as space and form, as place and pattern. Salvation is for all of creation. The liberating, healing, inclusive ministry of Christ takes place *in* and *for* creation.[66]

In this regard, she aptly challenges "the long antibody, antiphysical, antimatter tradition within Christianity."[67]

---

62 Gunton, "The Doctrine of Creation," p. 155.
63 Cf. Gunton, *The Triune Creator*, p. 142.
64 Gunton, "The Doctrine of Creation," p. 155.
65 McFague, *The Body of God*, p. 182.
66 McFague, *The Body of God*, p. 182.
67 McFague, *The Body of God*, p. 74.

146   *Transcendence and Spatiality of the Triune Creator*

The antithesis between God and the body is more philosophical than biblical. Under the influence of Greek dualism between soul and body, the early Christian church fathers developed the doctrine of the immortality of the soul. Augustine especially played an important role in this area of Christian theology. The Greeks held no belief in the resurrection of the body. The body was considered the grave of the soul. Death was thought to enable the emancipation of the soul from the shackles of the body. While admitting that there are controversies over the doctrine of the immortality of the soul in Reformed theology, Anthony Hoekema asserts that the central message of Christianity about the future of human beings is not the immortality of the soul but rather the bodily resurrection.[68]

Concerning the humanity of Jesus after his ascension, Reformed theologians, unlike Lutheran theologians, have asserted that the ascended Jesus still has his humanity, including a glorified body.[69] Dealing with God's temporal attributes, Gerald Bray indicates the following discrepancy between Luther and Calvin:

> According to the orthodox formulation, the person of the Son of God took on human nature (bound by time) which remains completely separate from his divine nature (not bound by time), in order to do his work within the time and space framework. The reconciliation of time and eternity occurred in the ascension which, scientifically speaking, is the greatest mystery in the life of Christ. When he ascended into heaven, Jesus took his manhood into God, thereby making it timeless. (Whether it was also made spaceless is a matter of controversy. Some, like Luther, have argued that it was, because time and space were usually held to be inseparable. Others, like Calvin, argued that it was not, on the ground that such a change would destroy the natural appearance of a human body.)[70]

---

68   Anthony A. Hoekema, *The Bible and the Future* (Grand Rapids: Eerdmans, 1979), p. 91.

69   Thomas F. Torrance is a representative Reformed theologian who emphasizes the vicarious humanity of Jesus. After citing Athanasius' saying that Jesus Christ exercised a two-fold ministry, Torrance avers that, "manward and Godward ministry are to be thought of as an inseparable whole in the oneness of our Lord's Person as God and Man, and as continuous throughout all the reconciling movement of his life to its culmination in his vicarious death and resurrection, *but also as extending after his ascension into his heavenly intercession as our High Priest and Advocate before the Face of the Father* (Italics added)" [*The Mediation of Christ* (Colorado Springs: Helmers &Howard, 1992), p. 73].

70   Bray, *The Doctrine of God*, p. 85.

Divine Spatiality   147

Thus, the discussions of the spatiality of God have more relevance for Reformed theologians than for other Christian traditions.

5. 'The Earth is God's'

From a Reformed perspective, William A. Dyrness proposes "A Theology of American Culture," in his book *The Earth is God's*. But, the book is not concerned not only with American culture, also with what can be said about general human existence in the world. In the preface of the book, Dyrness notes, "former ways of thinking about God – as sure knowledge or as a narrative of God's acts – have not provided the resources we need to address our corporate and embodied life in the world."[71] This criticism Dyrness directs at history and narrative theologians. He admits that history and story are certainly important tools. But, he says, they need correction and a deepening. As Dyrness states, "History and story are categories that readily incorporate time but not *space*. They imply movement, as in a melody, but they cannot fully comprehend the soil – the complexity in which this movement is rooted."[72] Concerning the importance of developing thought on space, he states,

> It is in space that we encounter the other person, with all of the strangeness that not only does not fit into our story but may challenge it in fundamental ways. As John MacMurray points out, our relation to others is primarily tactile, we encounter them but we do not comprehend them. So we must find ways to understand the physical nature of this encounter. For it is in space that our embodied practices are constructed, where we injure or heal the earth, where we shape objects and events that celebrate our deepest values.[73]

Thus Dyrness asserts that "Spatial and tactual dimensions must be added to our intellectual map and our spiritual journey."[74]

---

71   William A. Dyrness, *The Earth is God's: A Theology of American Culture* (Maryknoll, NY: Orbis Books, 1997), p. xi.
72   Dyrness, *The Earth is God's*, p. 12.
73   Dyrness, *The Earth is God's*, p. 13.
74   Dyrness, *The Earth is God's*, p. 160.

148   *Transcendence and Spatiality of the Triune Creator*

Dyrness seeks to present some theological tools to elaborate the fabric of our corporate life, with an eye to their relevance for addressing issues of clean water and ethnic conflict. According to him, some new ways of speaking of God that the most recent generation of theological scholarship has suggested, enable us to address these geopolitical concerns. Thus Dyrness says, "Grounded in the triune presence of God, we argue that the categories of relationship, agency, and embodiment, provide windows that helpfully illuminate our human life in the world."[75]

Pointing out that a gulf divides theology from our corporate life, Dyrness argues that God and our corporate lives should have much more to do with each other. The key to explaining this necessary relation between God and our corporate lives, according to Dyrness, is God's trinitarian character. On this note, Dyrness points to certain weakness in the theological formulations that we have inherited: "The language we have inherited to speak of God and the world has not been adequate to express this continuing relationship of God to the earth and its peoples [...] Even the more recent emphasis on understanding God's presence as 'story,' helpful in so many ways, reaches its limits when it comes to understanding our embodied stewardship of the created order."[76] Thus he proposes that our inherited language be complemented by three normative categories that grow out of God's own trinitarian presence and reality. He describes these three categories as follows:

> first, as God's nature is communal so our life in the world is invariably relational and communal; second, as God's presence is creative and active, so our human life in the world must be fundamentally understood in terms of the value of human agency; finally, as God has irrevocably identified with the created order both in his original creative purposes and in terms of the divine incarnation in that order, so the human and created order must be construed in terms of the value of embodiment.[77]

---

75   Dyrness, *The Earth is God's*, p. xii.
76   Dyrness, *The Earth is God's*, p. xiv.
77   Dyrness, *The Earth is God's*, p. xiv. Elsewhere he describes these three categories like this: "All three of these movements – God's own relationship and commitment to the creation; the work of God in making, sustaining, and renewing the creation; and the revelation of God's love and being in the human Jesus Christ, along with the continuing work of the Holy Spirit – are directed toward the perfection and glory of creation" (ibid., p. 16).

Divine Spatiality 149

In other words, these categories, according to Dyrness, are "pointers toward a theology of life in the world."[78]

Narrative theology is diachronic so that it can carry us through time and it does well in giving direction to our life in the world. Dyrness contends, however, our life has "a texture, a complexity that we dare not overlook" as well as "a *telos*, an end, that points to the final realization of God's glory in the new heaven and new earth." Thus he holds that "we must add to the diachronic method of narrative theology *synchronic* ways of thinking about our life in the world, ways that help us get a handle on our spatial and embodied lives in the here and now. For it is here in this complexity that we must seek to do the will of God, who called us out of darkness into light."[79] A major element of this complexity is "the way human and natural evil interrupts (and corrupts) the stories of our lives." Dyrness states the basic direction of his book as follows: "Theology must find ways to account for these 'storyless' spaces in our corporate and individual lives. Our trinitarian starting point and these pointers lead us to suggest three normative categories to use in articulating the theological reality of our life in the world."[80]

Dyrness maintains that there are two sides to the category of relationship which is rooted in the trinitarian basis: an active seeking and a passive receiving. He states that "relationship implies the ongoing commitment of God to the world." Rather than stressing God's immutability, Dyrness focuses on God's love: "God's being must be construed in the personal categories that scripture uses: God loves the world. Love and commitment imply an active seeking and making contact with what is there."[81] Thus, he states, "nothing is done, no relationship is formed, without this activity of commitment." He holds three aspects of human being's relationship: relationship with the living God, relationship with another human being, and relationship with the rest of creation. Then Dyrness indicates the passive side of relationship: "To be in relation is to receive from another what is there."[82] To explain this process, he quotes the saying of the Danish philosopher Knud Løgstrup, which in part reads as follows: "unless there is attunement, nourished by things in nature,

78   Dyrness, *The Earth is God's*, p. 16.
79   Dyrness, *The Earth is God's*, p. 16.
80   Dyrness, *The Earth is God's*, p. 17.
81   Dyrness, *The Earth is God's*, p. 18.
82   Dyrness, *The Earth is God's*, p. 18.

150 *Transcendence and Spatiality of the Triune Creator*

there would be no zest and energy for a single life manifestation." Thus Dyrness asserts that "reception cannot exist without this attunement, which reflects the created coherence of the world."[83] "This conscious and sustained openness to impression makes possible life together," he writes; "it is the ground of worship and art." Hence, for Dyrness, "relationship is both an active seeking and a no less active receiving." He argues that "our relationship with another, rather than constraining us (though it can appear to do this) is potentially expressive of all the interests, purposes, and feelings that make us what we are, and, in their intentionality, echo God's own commitment to the world."[84]

Agency is the second category of Dyrness. In the Western philosophical tradition, Dyrness maintains, "what happens, what is done, seems somehow less important than what we know (for sure)." In the Biblical Theology Movement of the 20th century, however, "the acts of God are fundamental to our knowledge of God; indeed, they have a certain priority." For Dyrness, thus, "our knowing is realized and verified – is often catalyzed – in our physical and active life in the world."[85] The emphasis on the human action, according to him, has "the twin dangers of unthinking and possibly destructive activism." But, Dyrness says, "as human beings, agency, often grounded in restless longing and loneliness, is constitutive of our life in the world."[86] Quoting LaCugna, he asserts that "God chooses freely to act, and this acting becomes the basis of all human agency, which is always and inevitably contextual. And as God did not reside in isolation, so human agency is also directed outward toward another."[87]

Embodiment is a third category that is necessary to our life before God: "Human life is necessarily a physical and bodily life." "Far from being a barrier to life with God," Dyrness comments, "bodily life in scripture has become the privileged vehicle for that life [...] [H]uman life is inevitably embodied and thus engaged with the physical world."[88] Referring to the incarnation of the Son, he states the following: "Just as his bodily presence becomes the vehicle of his saving work, so the Reign

83   Dyrness, *The Earth is God's*, p. 19.
84   Dyrness, *The Earth is God's*, p. 19.
85   Dyrness, *The Earth is God's*, p. 20.
86   Dyrness, *The Earth is God's*, p. 20.
87   Dyrness, *The Earth is God's*, p. 21.
88   Dyrness, *The Earth is God's*, p. 22f.

Divine Spatiality   151

that he introduces makes no distinction between the spirit and matter."[89]
After arguing that the nature of our salvation is bodily, Dyrness maintains that "life with God in heaven will not be disembodied but will involve a non-erotic and joyous bodily state."[90] Finally, he holds that human life is embodied worship. We worship the triune God in truth and
spirit (Jn. 4:24). But our worship is inevitably an embodied activity. In
this world, we worship God with our body and in our body. In the coming world, we will also worship God with our body and in our body. We
believe in the bodily resurrection.

6.   Divine Action and Embodiment

In his book, *God, Action, and Embodiment*, Thomas F. Tracy sets himself against the position of process theology that the world is the embodiment of God. Though he rejects Cartesian mind-body dualism,
Tracy does not agree with the analogy of divine embodiment in the
world. Tracy presents the concept of an agent as that of an individual
capable of intentionally regulating at least some of his/her own behavior.
Along with this concept of an agent, Tracy contends that "The concept of
mind as an active but immaterial entity promises to be useful in thinking
about God as an unembodied agent active in or upon the world."[91]

In Cartesian mind-body dualism, Tracy states, "the distinction between mind and body is radical": Whereas the essential characteristic of

89   Dyrness, *The Earth is God's*, p. 22.
90   Dyrness, *The Earth is God's*, p. 23. In this context Dyrness refers to the painting of
     the Italian painter Luca Signorelli in the San Brizio Chapel in the Orvieto Cathe-
     dral. Dyrness discusses five themes: creation, culture, ethnicity, ecology, and art.
     Robert I. Thompson, Jr., "The Earth is God's," *Missiology* 27/1 (Jan. 1999), p. 127,
     summarizes Dyrness as follows: "Creation comprises the limits that God has freely
     chosen, and culture is what humans fashion out of creation. Ethnicity is the identity
     that we obtain from what we make, or our culture. Ecology is identified with the
     challenges related to human stewardship of God's creation, whereas art embodies
     the human need to express in aesthetic form 'the deep splendor and agony of crea-
     tion' (135)."
91   Thomas F. Tracy, *God, Action, and Embodiment* (Grand Rapids: Eerdmans, 1984),
     p. 43.

152   *Transcendence and Spatiality of the Triune Creator*

matter, or body, is extension in space, the essential characteristic of mind is "thought," broadly understood to embrace all the contents and operations of consciousness. Tracy asserts that "it is the distinction between substance that is thinking and unextended on the one hand, and substance that is extended and unthinking on the other hand."[92] In this scheme the relation between mind and body has been always the issue. Regarding this problem, Tracy holds, "[t]he human body may be moved either by the motions of matter or by the initiatives of mind." In this context, Tracy describes the weakness of dualism in these terms: "if a bodily movement results strictly from the interaction of material bodies, then it is a mere automation."[93] He asserts that the Cartesian dualism commits a category mistake by treating mind as 'a type of substance, or entity" rather than "a set of capacities and characteristics" of an agent. According to Tracy, this mistake can be avoided by restating the dualism as existing between "a mental agent and a bodily mechanism," rather than between mind and body. However, Tracy holds that, "this restated dualism will […] still face the difficulties associated with the way it sorts out and sets apart the mental and the physical in human life."[94]

Tracy proposes that the human subject is to be thought of as "psychophysical unit" rather than, as with Descartes, a mental substance. Since the concept "agent" is closely tied to the concept of a psychophysical unit, the rejection of dualism seems to suggest that "an agent logically must be a bodily agent."[95] Tracy explains as follows: "When we abandon the dualist's account of intentional action as a causal control of bodily events by mental substance, we can no longer claim a direct analogy between the intrapersonal relation of mind to body and the cosmological relation of God to creatures in divine action."[96] Therefore, in order to

---

92   Tracy, *God, Action, and Embodiment*, p. 47.
93   Tracy, *God, Action, and Embodiment*, p. 48.
94   Tracy, *God, Action, and Embodiment*, p. 60. Polkinghorne also comments on the weakness of Cartesian dualism as follows: "The fatal weakness of his system is just the unresolved 'causal joint' by means of which the human ghost within manipulates the machine that it inhabits. Appeal to the pineal gland as the seat of the soul was a pretty desperate attempt at a remedy. In the end, Cartesianism had to fall back on occasionalism – the divinely synchronized ticking of the mental and material clocks, so that God arranges for my hand to move when I will it to move" (Polkinghorne, *Science and Providence*, p. 18).
95   Tracy, *God, Action, and Embodiment*, p. 60f.
96   Tracy, *God, Action, and Embodiment*, p. 61.

Divine Spatiality

express his own dilemma, Tracy presents some questions: "Can we make sense of the notion of an immaterial divine agent acting in and upon our world? Must an agent be a body in order to act upon bodies?"[97] In a review of three books, including Tracy's *God, Action, and Embodiment*, David Chidester describes Tracy's problem and his suggested breakthrough as follows:

> The problem he encounters is: How is it possible to talk about personal characteristics without a body, since most terms we use to identify personality traits – knowing, speaking, desiring, acting, and so on – presupposes an embodied state. The author follows the track layed out by Ryle and Strawson to conclude that a disembodied mind is like a square circle; and he transfers the challenge of analytic philosophy, in dissolving the mind-body dualism in a philosophy of persons, to a theology of God as person.[98]

Along with analytic philosophy, for me, the important tool of Tracy's is narrative theology.[99]

Tracy introduces Kaufman's interpersonal model of transcendence. According to Tracy, Kaufman's strategy is "to locate some point within our experience at which the concept of transcendence already has a meaning, so that this experienced transcendence can serve as a model for our relation to that which transcends us ultimately." Kaufman argues, Tracy writes, "there are important respects in which a human self is beyond the reach of others' knowledge unless he chooses to make himself known."[100] While some critics have argued that this interpersonal tran-

---

97 Tracy, *God, Action, and Embodiment*, p. 61.

98 *Journal of Theology for Southern Africa* 50/1 (Mar. 1985), p. 69. Chidester reviews Richard Swinburn, *Faith and Reason* (Oxford: Clarendon Press, 1981) and Keith E. Yankell, *Christianity and Philosophy* (Grand Rapids: Eerdmans, 1984), along with Tracy's book.

99 Grenz and Olson, *20th-Century Theology*, p. 271, indicate some contribution of narrative theology: "The new conceptualization developed by narrative theology lifts the discussion beyond the purely temporal category that the theology of hope introduced and that formed the basis for liberation theology's return to immanence. The genius of narrative theology lies in its assertion that faith entails the joining of our personal stories with the transcendent/immanent story of a religious community and ultimately with the grand narrative of the divine action in the world." But narrative theology does not push the theological discussion beyond time to space (Cf. Dyrness, *The Earth is God's*, p. 12).

100 Tracy, *God, Action, and Embodiment*, p. 63. See Gordon D. Kaufman, *God the Problem* (Cambridge, MA: Harvard University Press, 1972).

154 *Transcendence and Spatiality of the Triune Creator*

scendence trades upon a "residually Cartesian" under-standing of persons, Tracy maintains that Kaufman is right in claiming that a non-dualistic interpersonal mode of transcendence can be developed.[101] Through this interpersonal model of God's transcendence, Tracy asserts, we can discard "the claim that intentional self-disclosure, the voluntary making public of what is essentially private, lies at the foundation of all communication between persons." He says, "Much of our self-disclosure is *not* deliberately chosen. Being known by others comes along with being a person in a community of persons."[102] Thus Tracy holds that "[m]ental substance, after all, is conjoined to a body in a human life and interacts with that body in mutual causal inter-dependence."[103] Tracy admits that the rejection of dualism clearly eliminates certain simple and appealing patterns of theological argument for the divine action. But the shift to a nondualist understanding of persons, he maintains, does not eliminate any form of reflection crucial to theism; rather, such a shift can still retain a legitimate frame of reference for God's personal agency.[104]

Tracy raises the issue of the identifiability of God. If God is under-stood as a disembodied being, he is not a psychophysical unit as human beings are. According to Tracy, spatio-temporal relations provide the fundamental network of identifying relations. If God is not a psycho-physical unit, Tracy asserts, God cannot appear as a subject of reference within the scheme of identifying relations of space and time: "God [...] will not be a demonstratively identifiable entity in space and time."[105] Instead of "an exclusively story-relative identification," Tracy suggests, "a modified form of story-relative identification (viz., one which in-cludes reference to particulars within our shared field of experience) does have a significant role to play in talk of God."[106] Thus, citing Straw-

---

101 Tracy, *God, Action, and Embodiment*, p. 64.
102 Tracy, *God, Action, and Embodiment*, p. 64.
103 Tracy, *God, Action, and Embodiment*, p. 71.
104 Tracy, *God, Action, and Embodiment*, p. 65.
105 Tracy, *God, Action, and Embodiment*, p. 75.
106 Tracy, *God, Action, and Embodiment*, p. 75. Dyrness explains Tracy's argument as follows: "Christian truth can be story bound without being story-relative [...] The Christian story cannot be evaded because it includes references to events that be-long to our shared spatio-temporal frame of reference (81). That is to say, the Christian story, in a unique way, makes claims about and implies connection with the shared spatial world that grounds our lives and cultures" (*The Earth is God's*, p. 10).

Divine Spatiality

son's argument that, even though spatio-temporal relations provide the fundamental network of identifying relations, the individuating relation need not in every case be a spatio-temporal relation, Tracy suggests that "God might be identified by indicating a unique relation that he has to our shared world of objects and events located in space and time."[107] Finally, Tracy argues:

> Perhaps enough has been said to indicate the possibilities and problems that will arise for an attempt to identify God as an agent who cannot be "placed" spatially. Such an agent can be identified as a referent of human speech, but that reference takes place within a theistic story that includes but goes beyond the particulars of our ordinary network of references. The story we tell about God as an agent of intentional actions sets him in a unique relation to events within our shared field of experience. It may be difficult to state truth conditions for this story, but if one operates within the context it establishes, then individuating reference to God is possible.[108]

Concerning this issue, however, Jantzen aptly criticizes the argument of Tracy as circular reasoning: "God's actions can only be identified within the context of the whole theistic story, yet the story itself takes as its foundation incidents which must be identifiable as divine actions."[109]

Even though Tracy does not refer to McFague, Peacocke, or Jantzen, he acknowledges, "It is not uncommon for theologians to call upon the relation of an agent to his body as a model for the relation of God to the world. One might, that is, suggest that the world be thought of as 'God's body.'"[110] Here Tracy mainly has in mind process theologians whose roots lie in the philosophy of Alfred North Whitehead. The model of God as embodied in the universe, according to Tracy, can take forms both of mind-body dualism and of non-dualism. However, Tracy states, a non-dualistic model of God's embodiment will produce a considerably

---

107 Tracy, *God, Action, and Embodiment*, p. 76f.

108 Tracy, *God, Action, and Embodiment*, p. 83.

109 *Journal of Theological Studies* 37/2 (Oct. 1986), p. 675.

110 Tracy, *God, Action, and Embodiment*, p. 111. He cites further references: Charles Hartshorne, *Man's Vision of God* (New York: Harper & Row, 1941) and *The Divine Relativity* (New York: Yale University Press, 1948). And he also refers to Schubert Ogden, *The Reality of God* (New York: Harper & Row, 1963) and James McClendon, "Can There Be Talk about God-and-the-World?" *Harvard Theological Review* 62 (1969): pp. 33–49 (ibid., p. 174).

156   *Transcendence and Spatiality of the Triune Creator*

more radical revision of traditional theism than would a dualistic development of the same suggestion.[111] He comments as follows:

> Given a non-dualist understanding of the bodily agent as a psychophysical unit, God cannot be said to exist as a distinct being apart from and over against the world. Just as the human agent is not to be identified with an immaterial substance that controls the body, so God is not to be thought of as a supernatural power who intervenes "from above" to act upon the physical world. If we refuse to picture the human agent as a "ghost in the machine," then we should no longer represent God as a "ghost in the universe."[112]

Thus Tracy suggests, "To say that the world is God's body is to say that the processes unfolding in the universe are the processes of God's life, that God does not exist except in and through these processes."[113] If "God's life be rooted in a basic pattern of activity that he does not choose or intentionally enact, but that establishes and limits his capacity for intentional action," Tracy holds, "this quite clearly presents us with a distinctly finite God."[114] Thus, Tracy suggests God should be thought of as a non-embodied being if he represents the perfection of agency.

Vincent Brümmer argues that Tracy's view seems to presuppose the incoherent view that a thorough and perceptive analysis of human agency can be used as a conceptual model for talking about divine action. According to Brümmer, Tracy argues, "human agency is limited by a bodily substructure, and since the scope of God's agency is unlimited, he cannot have a body." Against this negative view on the body, Brümmer presents a positive and active role of body in human agency: "Our bodies do not merely *limit* our abilities, they *enable* us to perform those actions of which we are capable." Thus Brümmer states, "Far from being capable of unlimited actions, a non-bodily agent would therefore be incapable of performing any actions at all – except maybe mental actions, or 'projects for thought that neither require nor become full-fledged bodily actions' […] Clearly a God who can do no more than that, can hardly be called a 'perfect agent.'"[115] Grace Jantzen shares this criticism. While she admits that in some sense our physical body limits our range of in-

---

111   Tracy, *God, Action, and Embodiment*, p. 111.
112   Tracy, *God, Action, and Embodiment*, p. 112.
113   Tracy, *God, Action, and Embodiment*, p. 112.
114   Tracy, *God, Action, and Embodiment*, p. 116.
115   V. Brümmer, *Religious Studies* 21/4 (Dec. 1985), p. 605.

Divine Spatiality 157

tentional activity, Jantzen concurs that our physical structure is what makes activity possible at all. Thus she comments as follows: "To argue that a non-embodied being represents the perfection of agency rather than its cancellation one would have to show *how* non-embodied agency could occur at all (which [...] Tracy repudiates as unnecessary) or at the very least identify some actions as unequivocally the actions of a non-embodied being which are possible precisely in virtue of his non-embodiment."[116]

Of interest from our point of view is Clark H. Pinnock's review of Grace Jantzen's *God's World, God's Body* along with Tracy's book. They are published at the same year (1984) and share non-dualist views on the mind-body question. But they differ from each other in the issue of God's embodiment. Pinnock acknowledges that Jantzen's conclusion is more radical than that of Tracy. He summarizes Jantzen's purpose in this way: "She points out the little-known fact that the Bible does not really teach incorporeality, and shows how certain of the early Fathers did not believe in it either. By attributing a body to God, she believes that we can enrich the claims of theism and make them more plausible."[117] As is well-known, Jantzen's position is that the whole universe is God's body. Pinnock asserts that, "She is even prepared to admit that because God is by nature embodied the creation must be and is everlasting." In comparison to the view of Jantzen, Pinnock presents Tracy's problem as follows: "How can we avoid mind-body dualism at the cosmic level unless we conclude with Jantzen (whose book he [Tracy] has not read) that the world is God's body?" Pinnock tackles Tracy's contention that "if God is perfect on the scale of agency, God must be disembodied in order to be free." Thus, Pinnock contends that Tracy sacrifices the meaningfulness of God's agency in order to preserve his freedom. This move, Pinnock concludes, is a mistake: "I am certain that Jantzen will feel that Tracy has gained nothing by this move. God's agency remains the problem which it was at the outset of his book."[118]

Taking and modifying Jantzen's view that "God's body, unlike our own, is completely subject to his control and utterly dependent on him,"

---

116 *Journal of Theological Studies* 37/2 (Oct. 1986), p. 676.
117 *Christian Scholar's Review* 14/4 (1985), p. 391f.
118 *Christian Scholar's Review*, p. 392.

158   *Transcendence and Spatiality of the Triune Creator*

Pinnock contends that Jantzen's position provides more common ground than might at first appear. He writes,

> But this is very like a doctrine of creation in which God spreads forth a world in which he can embody himself and his purposes. Surely we can have the point Jantzen is making and not fall into Tracy's disembodied agency if we say that God creates worlds *to be* his body, to be the theatre of his self-embodiment. God in his freedom creates worlds where his life can be worked out in a creaturely process which, praise God, includes ourselves. Thus we dwell in a world which emanated from God's intention and which is full of God's presence and informed by his gracious purposes.[119]

Consistent with this view we may speak of creation in Calvin's terminology, as the theater of God's glory. Since Tracy's book does not leave a positive role for creatures to sing and enact God's glory, this criticism of Pinnock against Tracy seems justified. The denial of the view that the world is the body of God, however, need not imply that we should despise material aspects of the creation, including the human body.

Tracy does not refuse all forms of God's embodiment. Only "if embodiment means limitation by a given, subintentional pattern of activity" does he deny the embodiment of God. But "if embodiment means that God's life takes the form of a unified organic process," Tracy affirms such a view. Tracy maintains that "God is an absolutely self-determining agent who continuously enacts his own bodily life."[120] Tracy distinguishes between this modified theology of divine embodiment and a more traditional theism as follows: "On the one hand, we say 'God enacts the world as his body,' but we add the proviso that this body must be understood as (at least in part) a society of creaturely agents in and through which the divine agent lives his life; on the other hand, we might say 'God creates a world that includes finite agents whose existence he sustains and whose lives he subtly shapes and guides.'"[121] I would pro-

---

119   *Christian Scholar's Review*, p. 392. As a better idea than that of either Tracy or Jantzen, Pinnock presents John Macquarrie's view: "God transcends the world as an artist does his painting, but is also immanent in his work, having put so much of himself into it. Though transcendent, God goes out from himself to posit a world derived from himself and of infinite value to him. Thus there is no star which does not shine for God, and no place where his will is not struggling to be done" (ibid.). This view is similar to Peacocke's musical analogy.

120   Tracy, *God, Action, and Embodiment*, p. 119.

121   Tracy, *God, Action, and Embodiment*, p. 120.

Divine Spatiality 159

pose that we adopt these two views together, in a position I will develop in the following section, and in a way that is consistent with Dyrness's third category. One contribution of Tracy's work is his presenting the possibility that God can act in the world, even being without body. Thus, Tracy does not seem to deny embodiment itself but he does avoid the panentheistic implication of process theology.

I have suggested the fallacies with saying that God is a body or that God has a body, and I have refuted the position that the world is the body of God. So, I would argue that God is not a bodily being. But he is the creator of the world, including space, time, and bodies. And in Reformed tradition this conclusion has important implications. One such implication is that God is concerned about our bodily life in the world. Although he does not exist as something created, nor as a bodily being, God can indeed be in the space and time of creation. How can God be present in space in a way unlike a creaturely occupant?

7. God Is in Space

Making use of the term "perichoresis," Torrance refers to the possibility of the space or place of God. He states that "'place' for God can only be defined by the communion of the Persons in the Divine life."[122] Though he does not develop the patristic concept of perichoresis in relation to the relation of God to space, Torrance fully deploys his view of perichoresis in his later work. Perichoresis, according to him, is used "to express something of the mystery of the Holy Trinity in respect of the coinherent way in which the Father, the Son and the Holy Spirit exist in one another and dwell in one another as one God, three Persons."[123]

In the development of his detailed views on God and space, Moltmann overcomes the limitations of Torrance. He proposes the concept of perichoresis as a kind of divine space, and then extends this divine space into the creaturely level. Unlike Karl Barth, he seeks to find the unity of the triune God not initially in his subjectivity and the sover-

122 Torrance, *Space, Time and Resurrection*, p. 131.
123 Torrance, *The Christian Doctrine of God*, p. 168.

160 *Transcendence and Spatiality of the Triune Creator*

eignty of his rule but in the unique, perfect, perichoretic fellowship of the Father, the Son and the Holy Spirit. Following Johannine theology, Moltmann takes "as archetype of all the relationships in creation and redemption that correspond to God, the reciprocal perichoresis of the Father and the Son and the Spirit (John 17.21)."[124]

I have already dealt with the position of Moltmann on "God is space." Criticizing his view, I have indicated that Moltmann identifies the divine perichoretic space with the triune God himself. Moltmann correctly points out that, according to Scripture, we can exist in the perichoretic space of the triune God – That is the meaning of the second dimension in Jesus' prayer: "that they may also be in us" (Jn. 17:21). Yet, Moltmann is incorrect to suggest that we literally exist in the triune God. Through the Spirit, I've contended, we can exist in the perichoretic space of God. And in this sense, we exist in the triune God. Thus, by means of the term "perichoresis," we can affirm not only the mutual incoherence of the three Persons of the Trinity but also our indwelling in the triune God. Then, is it not possible to affirm that God is in the creaturely world of space and time as suggested by our usage of the term perichoresis?

This possibility is in fact proposed by Gunton in his discussion of co-presence of divine and human in Jesus. He contends that, since absolute space is externally related to God, God's co-presence with a particular spatial being becomes problematic. Rather, according to Gunton, space is not to be seen in terms of exclusiveness (absolute space) but of inter-penetration (perichoresis). By this means, Gunton claims, we can con-ceive of God at once as creator of space and as able to relate himself internally to parts of it. This argument revisits the ancient polemics of Athanasius against Arius. For Gunton, as in Torrance, "everything de-pends upon how we understand space."[125] By using perichoresis, Gunton thus seems to propose a relative, or relational, concept of space, in oppo-sition to Newton's concept of absolute space.

Gunton connects this idea of perichoresis to the field theory of mod-ern physics: "What we have in Faraday is a kind of doctrine of the *peri-cohresis*, the interpenetration, of matter. As the three persons of the Trin-ity interpenetrate the being of the others, so it is with the matter of which

---

124 Moltmann, *God in Creation*, p. 258.
125 C. E. Gunton, *Yesterday and Today: A Study of Continuities in Christology* (Grand Rapids: Eerdmans, 1983), p. 120.

Divine Spatiality　　　　　　　　　　　　　　　　　　　　　161

the world is made."[126] Gunton understands the character of the universe as "a *perichoresis* of interrelated dynamic systems"[127] or "a system of fields of forces, interpenetrating and interacting."[128] Quoting Faraday's saying that "matter is not merely mutually penetrable, but each atom extends, so to say, throughout the whole of the solar system, yet always retaining its own center of force," Gunton contends that "there is a sense in which *everything is everywhere*."[129]

In relation to the doctrine of the Trinity, perichoresis refers to the mutual indwelling of three Persons. It was originally employed to speak of the mutual interpenetration of two natures in Christ.[130] Gunton, however, applies this conception to designate the intra-trinitarian relations in God to the world of creatures. Perichoresis not only involves the triune God but also creatures. Thus, by the Spirit, we creatures can dwell in the perichoretic space of the Trinity. If we can affirm that perichoresis is not only a kind of divine space but also a phenomenon of this world of space and time, the triune God can be said to be within the world of space and time. In other words, God's being in space of this world becomes possible as we abandon the concept of absolute space and adopt the perichoretic concept of space as a kind of relational space. In this sense, I maintain that, even though God does not have a body, he can be present in the space of this world. Furthermore, God can be co-present with the creatures in one place, or space. This idea is related to the divine attribute of omnipresence. One more thing that I want to indicate here is that, although he is in space or is present in this world, God is not limited by the space of this world. Put in more traditional terms, this statement means: God is present or immanent in this world. At the same time, however, God is transcendent over the world of space and time.

126　Gunton, "Relation and Relativity: The Trinity and the Created World," p. 95.
127　Gunton, "Relation and Relativity," p. 106.
128　Gunton, *Yesterday and Today*, p. 118. In this context, Gunton observes that, "the claim of Professor T. F. Torrance (1969) that patristic conceptions of cosmology are nearer to ours than is often assumed, comes into focus" (ibid.).
129　Gunton, *Yesterday and Today*, p. 117.
130　Torrance, *The Christian Doctrine of God*, p. 102, indicates that the term "perichoresis" was used first by Gregory Nazianzen in a verbal form to help express the way in which the divine and the human natures in the one Person of Christ in one another without the integrity of either being diminished by the presence of the other.

162   *Transcendence and Spatiality of the Triune Creator*

In his eschatology, Moltmann develops this idea of perichoresis more radically. He contends for the mutual interpenetration of God and the creatures at the eschaton. Moltmann understands the concept of mutual indwelling as mutual interpenetration. So he contends that "the concept of mutual interpenetration makes it possible to preserve both the unity and the difference of what is diverse in kind: God and human being, heaven and earth, person and nature, the spiritual and the sensuous."[131] In addition to the Jewish contention that "The Lord is the dwelling space of his world," following the exposition of Jewish and Christian Shekinah theology, Moltmann argues that creation is also destined to be the dwelling space for God.[132] As a result, Moltmann believes, God's being 'all in all' (1 Cor. 15.28) will come true at last.[133] This belief is frequently cited as Moltmann's panentheistic vision. Through this mutual indwelling of God and the world, a so-called "cosmic *communicatio idiomatum*, a communication of idioms" takes place. In describing this, Moltmann uses the scholastic phrase, "mutual participation in the attributes of the other": "Created beings participate in the divine attributes of eternity and omnipresence, just as the indwelling God has participated in their limited time and their restricted space, taking them upon himself."[134] Moltmann's vision is thus made possible with the help of the perichoretic concept of space. In this coming panentheistic state, God and the world are not separable but still mutually distinguishable. Through the communication of idioms, God and the world mutually participate in the attributes of the other. Though this eschatological vision is interesting in many ways, I think Moltmann pushes his idea too far. While we can admit the participation of creatures in the divine attributes of eternity and omnipresence, the participation of God in the limited time and the restricted space of creatures risks damage or loss to God himself.

The indwelling of God in this world is not exclusively an eschatological event. God already indwells the world of space and time. He is present in space of this world. Nevertheless, God is not limited within this creaturely space. He is transcendent over this world. For God is the creator of space and time. Schwarz's model of dimensional relationship between God and the world, which I described earlier, makes sense of

131   Moltmann, *The Coming of God*, p. 278.
132   Cf. Moltmann, *The Coming of God*, p. 307.
133   Moltmann, *The Trinity and the Kingdom*, p. 104f.
134   Moltmann, *The Coming of God*, p. 307.

Divine Spatiality 163

both divine transcendence and immanence alike: "God would be present in our dimension (i.e., four-dimensional space-time continuum) without being contained by it and he would transcend it without being absent."[135] Generally speaking, this model seems to be valuable. It effectively represents both the divine transcendence and immanence. But I have one reservation about this dimensional model of relating God and the world. This model does not seem to allow a space for the world apart from God. Thus, in some sense, it slides into the panentheistic idea that the world is within God. If we acknowledge that there is a limitation of figuring God's relation to the world visually, however, this is not a major problem.

One more thing of value from Schwarz's article is his reference to the quality of God's presence: "God is not uniformly present in the world. Though present extensively everywhere, God is not present with the same intensity. Thus we can speak of the distant God and the God close at hand."[136] Expressions such as "God has left us" and "God is coming to us," are not, for Schwarz, just figurative sayings, but are statements which express a different quality of God's presence. God's presence is understood in terms of his activity. Yet, Gerald Bray opposes Schwarz's view, on the grounds that, first, "God is fully consistent in his omnipresence, *i.e.* there are no parts of him which have a greater concentration of divinity than others, and second, that he does not extend himself in space, like a smell or an oilslick. This last observation is very important because of the belief held by process theologians, that God is expanding along with his universe. God cannot expand because he already fills everything in the total consistency of his being."[137]

Introducing Augustine's view on God and space, Thomas Tracy states, "The denial that God occupies *a* space (i.e., a body), coupled with the assertion that God is always present, suggested to Augustine the idea of a being uniformly distributed throughout the universe."[138] But indicating that this view entails that "the body of an elephant would contain more of [God] than the body of a sparrow to the extent that it is larger and occupies more space," Tracy argues that Augustine has not overcome spatial thinking in his effort to envision omnipresence. Augustine

135 Schwarz, "God's Place in a Space Age," p. 367.
136 Schwarz, "God's Place in a Space Age," p. 361.
137 Bray, *The Doctrine of God*, p. 97.
138 Tracy, *God, Action and Embodiment*, p. vii.

164 *Transcendence and Spatiality of the Triune Creator*

could think of God only by using spatial metaphors which he then carefully qualified. Thus Tracy concludes that, "Augustine artfully 'breaks' his spatial metaphor in just the way necessary to help us toward the idea of a being who is fully present in every place yet located in none."[139]

8. Conclusion

This chapter has examined various options for speaking about God's spatiality. Moltmann's contention that God is space is a mistake that identifies the divine perichoretic space with the triune God himself. We can exist in the divine space but cannot be within God spatially and literally. Through the Spirit, we can dwell in the triune God. Thus, the option that God is space is to be understood metaphorically. Otherwise, this option is to be discarded. Concerning Newton's contention that space is the *sensorium Dei*, there is controversy. If he is describing the absolute space as a kind of divine sense organ, this contention connotes a panentheistic idea. But if his contention is that God uses space as the medium of his creative presence in finite places, the notion is acceptable. In this case, however, we need some concept of space other than that of absolute space. I have supported the position that God has his own space other than that of creatures. I have appealed to the dimensional difference between God and the world. If God has his own space, his space is different from that of creatures. God is not a bodily being. Yet, God is not hostile to body, but rather has a positive relation to the material world. Because he is the creator of the world, God is concerned about our bodily life in this world. To occupy a place or space, creatures need a kind of material body. Without a body, however, God can still exist in space and interact internally with the universe. With the term "perichoresis," we may designate the way in which the three Persons of the Triune God mutually indwell one another and the way we also indwell the triune God. Finally, I have argued for the view that God is in the space of this world. The crucial point here is the transcendence of God over the world of space and time. Even though he is present in the limited space

139 Tracy, *God, Action and Embodiment*, p. viii.

Divine Spatiality   165

of creatures, God should be still transcendent over our space. A dimensional model of relating God and the world represents an example of reasonable discourse about God's immanence or presence and transcendence.

Dyrness indicates that our bodily life on the earth is the embodied worship of God. Dyrness emphasizes our bodily life by the three categories which are grounded in the triune God's presence in the world: relationship, agency, and embodiment. Spatial and tactual dimensions must be added to our intellectual map and our spiritual journey. Dyrness points out two very important implications of his argument:

> First, the truly Christian mode of knowing, as of living, is not an abstraction from the world and a hankering after a mystical peak experience, but a concrete involvement and self-giving in the world's God-giving physicality. Life involves a careful seeing in the midst of life, a collecting from it of the gifts God has put there and given meaning by the divine presence [...] Second, then, human life points toward the future God is planning for it, in which God and creation are perfectly integrated and God's trinitarian life perfectly displayed.[140]

Tracy opposes the embodiment of God. However, his main target is not Dyrness's view on embodiment, but rather the view of process theologians, that the world is the body of God. Tracy does affirm some kind of divine embodiment. In the next chapter, deploying fully these three categories of Dyrness, I will present some positive meanings of God's spatiality. To be present in the world, God needs some kind of spatiality.

---

140 Dyrness, *The Earth is God's*, p. 161f.

Chapter VI

The Presence of the Triune God

1. Introduction

This chapter will present three ways of approaching God's spatiality. I will argue that it is necessary to speak of divine spatiality if we are to speak about the presence of the triune God. The position presented here is that not only is the triune God relational intra-trinitarianly, but also he has a real relationship with this world. God is not only the creator of this world. He now acts in it. God has embodied his glory in various ways in this world. Accordingly, we may employ Dyrness's categories, relationship, agency, and embodiment, as means by which to express the presence of the triune God in terms of divine spatiality.

This argument builds on recent studies of the Trinity. One of the characteristics of modern theology is the renaissance of the trinitarian theology. Since Barth and Rahner, almost every theologian has been involved in the discussion of the doctrine of the Trinity. Through the theological works of various theologians of the last century, the doctrine of the Trinity has emerged as more than just one doctrine among other doctrines; it is one of the proper characteristics which distinguishes Christianity from other theistic religions, such as Judaism or Islam. This doctrine of the Trinity has been developed as a theology of relationship. It is by means of such a dynamic trinitarian structure that we can express God's relationship, or relationality, without denying his transcendence. The triune God has himself embodied in various ways. The theology of embodiment is christological or pneumatological. Christ and the Spirit are seen as the two hands of God. Thus the theology of embodiment should be trinitarian.

Another characteristic of the modern theology relevant to our study is its openness to dialogue between theology and science. In this regard, Torrance, Pannenberg, and Moltmann are representative theologians,

168 *Transcendence and Spatiality of the Triune Creator*

who have themselves engaged in discourse with scientists. In addition to them, there have emerged some "scientists as theologians." Ian Barbour, Arthur Peacocke and John Polkinghorne are representative of such scientists as theologians.[1] For them, science and theology are not based on different, or independent, languages. Science and theology need not be in a conflict, indeed they should be mutually influential. Science and theology must influence each other. In the world view of Newtonianism or Laplacian determinism, there is no "place" for God in the universe. But the more recent model of the universe, as depicted by science, has rapidly changed. Laplacian determinism is no longer tenable. The universe is seen as open-ended. This change in the scientific atmosphere invites new explorations of God's action in the universe. If the universe were closed and causally determined, since God should not be considered a cause like other creatures, God could not act as an agent in the world. As a solution to this intellectual dilemma, some theologians, such as Schleiermacher, have presented the option that, while God cannot act in history, God does enact of the history of the world. But this position is in fact not distinguishable from deism. This view does not admit God's place in the universe. In other words, it does not acknowledge space, room or gaps for God's action in this world. However, contemporary natural science finally has come to recognize that it is not possible to explain various phenomena of this world completely, without acknowledging indeterministic gaps. Within the universe, there are gaps. These gaps are not just epistemological, as in the view of the God of the gaps.[2] Thus, God can freely interact with or act in this universe.

Section 2, below, deals with the category of relationship. The triune God exists immanently in relation, prior to having a relation to the world. Augustine argued for the relationship of God. The contemporary theological discussion includes the history of the world into God's intra-

---

1   Cf. John Polkinghorne, *Scientists as Theologians: A Comparison of the Writings of Ian Barbour, Arthur Peacocke and John Polkinghorne* (London: SPCK, 1996).

2   In his article "Particular Providence and the God of the Gaps," Thomas F. Tracy suggests that "God would be the creator not only of natural law but also of the indeterministic gaps through which the world remains open to possibilities not exhaustively specified by its past" [*Chaos and Complexity: Scientific Perspectives on Divine Action*, eds R. J. Russell, N. Murphy and A. R. Peacocke (Vatican City State: Vatican Observatory Publications/Berkeley, CA: The Center for Theology and the Natural Sciences, 1997, 2nd Ed.), p. 294]. Following Tracy, I use the term "gaps" in a neutral sense. I do not support the God of gaps.

The Presence of the Triune God 169

trinitarian relation. By means of the Christ events – including the incarnation, the cross, the resurrection of the Son – the history of this world is neither contingent nor incidental, but has an intra-trinitarian significance. The triune God is not an isolated being but a relational being and has a kind of relationality. To speak of this relationship requires speaking of God's spatiality. By the term "God's spatiality," thus, I am emphasizing that the triune God exists in relations, or has relationality.

Section 3 examines Dyrness's second category, agency. The development of modern natural sciences narrowed down the scope of God's action and finally exiled him out of the universe. Laplace claimed that he did not need the hypothesis of God in his system. The universe in this world-view is a closed system. There is no space or room for God's action. To speak of God's intervention in this kind of world would be to introduce an irreconcilable contradiction. Thus, Schleiermacher, Kaufman, and Wiles contend for God's enactment *of* history but deny that God enacts *in* history. The universe as a closed system, however, is now refuted by scientists themselves. Hence there are space, room, or "gaps" for God's action in this world. Developments in such areas as biology, quantum theory and chaos theory are held to signal conceptions of further openness in the shape of reality and gaps for God's action. Peacocke, for example, describes chance as God's radar. He supposes that God interacts with the world through top-down causation. Unlike Peacocke, in addition to top-down causation, Murphy and Polkinghorne complement God's action in terms of a bottom-up causation. Murphy suggests that quantum indeterminacy presents us with proof of room for God's action. Polkinghorne opts for chaos theory as consistent with the possibility of divine action. The God who is active in this world, or God as an agent, has space, room, or gaps for his own activities within the world, or universe. This is the second meaning of God's spatiality. Discussions in these areas refer little to the doctrine of the Trinity. In this sense, these discussions are not proper to Christianity but belong to theism in general. For they are not grounded on divine revelations but on scientific discoveries and theories.

Section 4 focuses on Dyrness's third category, embodiment. Here the purpose is to find some ways to speak of how God has embodied himself. This discussion will reflect on the following points. As in Calvin, creation is the theater for God's glory. In the incarnation of the Son, God embodied himself in the world of space and time. The church, as the

170   *Transcendence and Spatiality of the Triune Creator*

people of God, of both the Old and New Testaments comprises the earthly-historical form of God's own existence. The Eucharist is the visual and material means of God's invisible grace. Christ is present in the Eucharist. The new creation is the work of making the creation a suitable vehicle for God's glory. Ultimately, this new creation will be an embodied event. In some sense, it has already started. God uses space, time and bodies as vehicles of his grace. By means of them, God expresses his goodness and love for his creatures. Thus, although we may refute the idea of the world as the body of God, we may still acknowledge a kind of God's embodiment. In these processes of embodiment, God uses material means, including the human body, as mediums of his presence.

2. Relationship

Dyrness's first category, that is, relationship, suggests that we may speak of God's spatiality in terms of divine relationality. This view can be connected with the recent development of the doctrine of the Trinity. One of the most prominent characteristics of the last century of theological reflection is the renaissance of the doctrine of the Trinity. Unlike in Kant, the doctrine of the Trinity is no longer a speculative dogma irrelevant to the church and the practical life of Christians. Through the theological works of various theologians of the last century, the doctrine of the Trinity has become neither an addendum to dogmatics (Schleiermacher) nor a dogma dealt with after the unity of God (Aquinas), but one of the characteristics which distinguish Christianity from other theism such as Judaism and Islam (subsection 1). The renaissance of the doctrine of the Trinity can be in one sense characterized as the emergence of the theology of relationship. Behind this development, there is a new understanding of the person as defined by dynamic relations. The person is no longer viewed as a windowless monad, in the Leibnizian sense, but rather is understood in relation to others. The triune God likewise has been reexplained as existing immanently in relations, even prior to God's relation to the world. Actually Augustine anticipated this discussion early on, as he discussed the relationship of God. The contemporary theological dis-

The Presence of the Triune God

cussion includes the history of the world as viewed in terms of God's intra-trinitarian relations. By means of the Christ events (i.e., incarnation, cross, and resurrection), the history of this world is neither contingent nor incidental but has an intra-trinitarian significance. The triune God is not an isolated being but a relational being and has a kind of relationality in Ted Peters' term (subsection 2). This relationality of God is deeply developed by LaCugna and Pannenberg. Yet, more needs to be said about how this relationality of God implies God's own spatiality. With the term "God's spatiality," I am emphasizing that the triune God exists in relations, or has a relationality. This discussion accordingly points to the importance of the immanent Trinity, while also denying the concept of the absolute independence of creatures (subsection 3).

2.1 The Renaissance of Trinitarian Theology

In the introduction to the book, *Trinitarian Theology Today*, Christoph Schwöbel indicates one of the most interesting developments in systematic theology in recent years. It is a renewed interest in the doctrine of the Trinity and its implications for various aspects of Christian theology.[3] We greet the age of "the renaissance of trinitarian theology."[4] The doc-

---

3   C. Schwöbel, "The Renaissance of Trinitarian Theology: Reasons, Problems and Tasks," C. Schwöbel, ed., *Trinitarian Theology Today: Essays on Divine Being and Acts* (Edinburgh: T&T Clark, 1995), p. 1. He describes this changed theological atmosphere as follows: "While at the beginning of this period it still seemed necessary to lament the neglect of trinitarian reflection in modern theology and to offer apologies for engaging with such allegedly remote and speculative issues, both, lamentation and apologies, would seem to be out of place in today's theological situation" (ibid.).

4   Catherine LaCugna also indicates this renaissance of the doctrine of the Trinity: "In all traditions today a renaissance of the doctrine of the Trinity is taking place, in Orthodox theology through the work of John Zizioulas, Christos Yannaras, and Stanley Harakas; in Protestant theology through the work of Eberhard Jüngel, Jürgen Moltmann, and Wolfhart Pannenberg; in Catholic theology through the work of Karl Rahner, Walter Kasper, Piet Schoonenberg, and others" [C. LaCugna, *God for Us: The Trinity and Christian Life* (New York: HarperSanFrancisco, 1991), p. 144].

172   *Transcendence and Spatiality of the Triune Creator*

trine of the Trinity is not just a doctrine among other doctrines but has become "a structural principle of theology."[5]

Catherine LaCugna introduces two representative forms of the neglect of trinitarian theology. First, she indicates the anti-trinitarian movements beginning in the sixteenth century. They rejected the doctrine of the Trinity "because of its lack of scriptural basis, its contrariness to reason, and its irrelevance to the practice of faith."[6] LaCugna states, "The so-called 'Father of modern theology', Friedrich Schleiermacher, was skeptical that the speculative doctrine of the Trinity could serve as anything more than an appendix to dogmatics. In this he represented the whole ethos of the Enlightenment."[7] Schleiermacher conceived of the

---

5   Concerning the position of the Trinity in the theology of Pannenberg, Burhenn claims that "the Trinity cannot function for Pannenberg, as it does for Barth, as a structural principle of theology" [Herbert Burhenn, "Pannenberg's doctrine of God," *Scottish Journal of Theology* 28:6 (1975), p. 536]. However, this assertion is superficial. In contrast to Burhenn, Galloway pointed out the significance of the Trinity in the system of Pannenberg: "Without this ontological basis [the trinitarian ontological basis] his doctrine of history as revelation would be mere metaphor" [Allan D. Galloway, *Wolfhart Pannenberg* (London: Allen & Unwin, 1973), p. 112]. Timothy Bradshaw, *Trinity and Ontology: A Comparative Study of the Theologies of Karl Barth and Wolfhart Pannenberg* (Lewiston, NY: The Edwin Mellen Press, 1988), p. 139, also interprets Pannenberg as a trinitarian theologian like Barth: "Pannenberg is another trinitarian theologian. He seeks to avoid what he considers to be Barth's errors, notably those stemming from dualistic tendencies, and yet consciously to appropriate the fundamental insights developed by Barth."

6   LaCugna, *God for Us*, p. 144. Concerning the practical dimension of the Doctrine of the Trinity, she asserts that, "The doctrine of the Trinity is ultimately a practical doctrine with radical consequences for Christian life [...] Because of God's outreach to the creature, God is said to be essentially relational, ecstatic, fecund, alive as passionate love. Divine life is therefore also *our* life" (ibid., p. 1).

7   LaCugna, *God for Us*, p. 144. Ted Peters rightly indicates the influence of Immanuel Kant on the doctrine of the Trinity as follows: "The inner dynamics of the divine being belong to the noumenal realm, whereas human cognition is strictly limited to the phenomenal realm. Kant opened the nineteenth century by diking off the flow of trinitarian speculation" [T. Peters, *GOD as Trinity: Relationality and Temporality in Divine Life* (Louisville, KY: Westminster/John Knox Press, 1993), p. 83]. One of the following three responses to Kant, according to Peters, is Schleiermacher, "who denies that the doctrine of the Trinity is essential to the expression of the Christian faith, thereby relegating it to the status of a second-rank doctrine" (ibid.). Moltmann also states that Schleiermacher conceived of the doctrine of the Trinity as secondary and understood Christianity as a 'monotheistic mode of belief' (Moltmann, *The Trinity and the Kingdom*, p. 3).

The Presence of the Triune God

173

doctrine of the Trinity as an inappropriate attempt to speculate on the inner life of God. Thus, LaCugna contends that in this view, "God's relationship to the creature entails no distinction or relation 'within' God, because the Absolute Subject cannot be partitioned. Schleiermacher was of the opinion that one could give a better account of the experience of Christian faith by excluding the dogmatic overlay of trinitarian doctrine."[8]

As the second form of the neglect of the doctrine of the Trinity, La-Cugna points out the separation of the theology of God from the economy of salvation by treating *De Deo Uno* and *De Deo Trino* as discrete treatises in medieval Latin theology. Following Augustine, medieval Latin theology reached its high point in Thomas Aquinas. In the *Summa Theologia* of Aquinas, "Theology of the triune God appeared to be added on to consideration of the one God. Unlike the metaphysics of the economy worked out by the Greek Fathers, scholasticism produced a metaphysics of the inner life of God."[9] LaCugna asserts that, "The much more significant structural feature of the *Summa* is its starting point with the divine essence, explored apart from its existence in triune personhood. The way for this had been prepared by Augustine, but Thomas' innovation was to use the metaphysics of Aristotle as the basis for his theology."[10]

According to LaCugna, however, there is a more basic reason for the marginalization of the doctrine of the Trinity that even underlies the failings of these periods. In part I of her book, she describes the emergence and defeat of the doctrine of the Trinity. By the triumph of the doctrine of the Trinity, she means the fact that "there is a possibility of a trinitarian theology that will be christological and pneumatological and therefore inherently related to Christian life and praxis."[11] In contrast to this possibility, the doctrine of the Trinity had been defeated since the

---

8 LaCugna, *God for Us*, p. 251.
9 LaCugna, *God for Us*, p. 10f. Similarly, she writes, "The essential attributes of God's nature are covered in *De Deo Uno*, and the tract on the Trinity treats only the formal aspects of the divine persons and relations. Trinitarian theology thus appears to be added on to consideration of the one God. Moreover, the treatise on the Trinity is unrelated to the doctrine of creation and, indeed, unrelated to the rest of theology" (ibid., p. 6).
10 LaCugna, *God for Us*, p. 147.
11 LaCugna, *God for Us*, p. 13.

174   *Transcendence and Spatiality of the Triune Creator*

Council of Nicaea (AD 325). As she explains this: "The diversity and uniqueness of the divine persons within the economy of redemption faded into the background, and the centrality of christology, soteriology, and pneumatology in the theology of God was diminished. Hence the defeat of the doctrine of the Trinity."[12] She argues that the basic root cause of the neglect of the doctrine of the Trinity is the split that occurred at Nicea between *theologia* and *oikonomia*. She analyzes the so-called Arian controversy as the background of the Council of Nicaea. For Arius, she states, "even though God *(ho theos)* cannot suffer, still God suffers in the person of the Logos, though it is a lesser God who suffers. In this respect Arius disjoined *theologia* from *oikonomia*."[13] For the pro-Nicenes, however, "Christ was not simply a bridge between the eternal God and human history, but the coming of *very God* into the world. Christ is the economy of God. In this respect they operated out of a correlation between *oikonomia* and *theologia*."[14] But, in opposition to Arius, who asserted that "the subordination of Christ to God according to the economy *(kat' oikonomian)* implied subordination at the level of the God's being *(kata theologian)*," the Nicene fathers were preoccupied with the eternal ground of *oikonomia*. After Nicaea, therefore, LaCugna holds, "the relationship among the divine persons was pursued in its theological and ontological, rather than its economic, dimensions. By the end of the fourth century, the self-relatedness of God was at the forefront of theological reflection, taking precedence over God's relationship to us in the economy of incarnation and deification."[15] Against this tradition, she contends that "*theologia* and *oikonomia*, the mystery of God and the mystery of salvation, are inseparable [...] The unity of *theologia* and *oikonomia* shows that the fundamental issue in trinitarian theology is not the inner workings of the 'immanent' Trinity, but *the question of how the*

---

12   LaCugna, *God for Us*, p. 9. As another scholar who uses the metaphor of defeat, she refers to D. Wendebourg "From the Cappadocian Fathers to Gregory Palamas: The Defeat of Trnitarian Theology," *St Patr* XVII/1 (1982), pp. 194-97 (ibid., 18).

13   LaCugna, *God for Us*, p. 35.

14   LaCugna, *God for Us*, p. 35.

15   LaCugna, *God for Us*, p. 54. Peters presents LaCugna's explanation as follows: "What happened in the process of writing the Nicene Creed was the separation of soteriology from the doctrine of God, so that *theologia* came to refer to the inner workings of the divine life apart from the work of salvation. The intradivine relations of the three persons lost their link to God's activity in the world" (Peters, *GOD as Trinity*, p. 123).

The Presence of the Triune God

trinitarian pattern of salvation history is to be correlated with the eternal being of God."[16]

The renaissance of trinitarian theology is related to concerns raised about the relationship of God to himself and his creation. LaCugna describes trinitarian theology as "par excellence a theology of relationship." She states that it "explores the mysteries of love, relationship, personhood and communion within the framework of God's self-revelation in the person of Christ and the activity of the Spirit."[17] It is in these ways that trinitarian theology is theology of relationship.

2.2 Theology of Relationship

Ted Peters indicates two moments in recent Trinity talk, which have begun moving us toward a much more satisfactory understanding of the doctrine of the Trinity. "The first is Karl Barth's insistence that the incarnate history of Jesus Christ incorporates temporality into the divine experience."[18] The second is so-called Rahner's rule or axiom, that the economic Trinity is the immanent Trinity, and vice versa. Peters seeks to go a bit further than Rahner might have intended. He writes, "the loving relationship between the Father and the Son within the Trinity *is* the loving relationship between the Father and Jesus. Relationships require otherness. Love posits and promotes otherness. What binds others in

---

16 LaCugna, *God for Us*, p. 4. Beginning with examining closely the transcripts of the core conversation that begins with Karl Barth and Karl Rahner, Peters, in *GOD as Trinity*, deploys the trinitarian thoughts of Eberhard Jüngel, Jürgen Moltmann, Robert Jenson, Wolfhart Pannenberg, and Catherine LaCugna. In addition to this discussion, he also includes topics such as the becoming of God of process theology, sexism in Trintarian language of feminist theology, and divine and human community of liberation theology. He claims that, "What one hears here is that through the incarnation God has experienced otherness, and through the power of the spirit God is experiencing an integrating wholeness. The result is that relationality has become a key theme in today's Trinity talk" (Peters, *GOD as Trinity*, p. 7). Peters states that, "The relational character of the divine life is the most salient concern of the work of Catherine Mowry LaCugna, for example, whose work places her much more squarely in the Barth and especially the Rahner tradition" (ibid., p. 122).

17 LaCugna, *God for Us*, p. 1.

18 Peters, *GOD as Trinity*, p. 21.

176   *Transcendence and Spatiality of the Triune Creator*

love is spirit."[19] The focus of the recent Trinity discussion, according to Peters, is on God's relationality. Peters asserts, "What the idea of the Trinity does is impute the quality of relationality to the internal life of God as well as to God's relationship to the world. This makes Trinity talk conceptually viable in a modern and emerging postmodern culture where relationality is integral to our understanding of reality."[20]

Among Aristotle's ten categories of being, unlike substance, the category of relation is an accident with the other eight categories that include quantity, quality, place, time, posture, acting, and being acted on.[21] Colin Gunton maintains that, "In Aristotle, and certainly in logic until the time of Kant, relation is subordinate to substance. Relations are what take place or subsist between substances that are prior to them: something first exists, and then enters or finds itself in relation to other things, which may change its accidents, but not what it really is (short of destroying it)."[22] Like LaCugna and Gunton, Peters questions a substantialist presumption, i.e., the distinction between absolute essence and relational attributes. Regarding substantialist metaphysics, he says, "The essence of an entity is absolute, remaining unchanged if identity is to be maintained. Relationality takes place through the attributes. What could not be countenanced is the notion that the divine essence is contingent upon the relational dimensions of its being."[23]

Behind the discussion of this trinitarian relationship, there is a new understanding of the person. Peters comments on this problem: "the modern notion of person seems to import too much individuality and independence for the traditional Trinitarian formula to work. But there is also emerging a postmodern point of view, one that emphasizes relationality. Its relevant axiom is that persons are always interpersonal. No one can be personal except in relation with other persons."[24] Thus, for Peters,

19   Peters, *GOD as Trinity*, p. 22.
20   Peters, *GOD as Trinity*, p. 30.
21   LaCugna, *God for Us*, p. 58.
22   C. E. Gunton, "Relation and Relativity: The Trinity and the Created World," ed. Christoph Schwöbel, *Trinitarian Theology Today*, p. 106.
23   Peters, *GOD as Trinity*, p. 31. According to Peters, this understanding of God's being has run into major obstacles in modern thought: "first, the denial that we could know God in the Godself, and second, the apparent incompatibility of an eternal unchanging God with the biblical view of a God in relationship to a world he loves" (ibid., p. 31f).
24   Peters, *GOD as Trinity*, p. 35.

The Presence of the Triune God

"person" should be thought of in essentially relational terms. Going one step further, Gunton asserts, "The persons are not persons who then enter into relations, but are mutually constituted, made what they are, by virtue of their relations to one another."[25] Gunton connects this understanding of "persons as relations" to Einstein's relativity principle. He conceives of relativity theory as "the application of some such conception of relations as the trinitarian one just outlined to the universe."[26] Thus, Gunton writes, "There are no unchanging substances which enter into relations – as on the view of Aristotle and Newton alike – but the whole universe becomes conceivable as a dynamic structure of fields of force in mutually constitutive relations."[27]

Discussing Augustine's doctrine of the Trinity, Sarah Lancaster states the following: "the terms *Father*, *Son*, and *Spirit* (and he [Augustine] later adds *Word* and *image*) are said with reference to each other. The Father is the Father of the Son, the Son is the Son of the Father, and so on. These relations are not accidents because they do not change, and yet they are not substance terms because they are not said simply with reference to self. The Father cannot be Father with reference to self, but only with reference to the Son."[28] Examining the Cappadocians' doctrine of the Trinity, LaCugna also contends that, "The definition of divine person as relation of origin means that to be a person is to be defined by where a person comes from; *what a person is in itself or by itself cannot be determined*."[29] Thus, she claims, "*it is impossible to think of the divine person in an entirely abstract way* disconnected from their presence in salvation history because it is only through the Son and Spirit that the unknowable God (Father) who dwells in light inaccessible is revealed to us."[30] Under the influence of Augustine, however, according to LaCugna, "the sharpened distinction between the triune God of salvation history and the Trinity of persons within God" took place in the West. It drastically transformed the direction and substance of future Christian theology. She describes this drastic transformation as follows:

---

25 Gunton, "Relation and Relativity," p. 106f.
26 Gunton, "Relation and Relativity," p. 107.
27 Gunton, "Relation and Relativity," p. 107.
28 Sarah H. Lancaster, "Divine Relation of the Trinity: Augustine's Answer to Arianism," in *Calvin Theological Journal* 34/2 (Nov. 1999), p. 334.
29 LaCugna, *God for Us*, p. 69.
30 LaCugna, *God for Us*, p. 70.

178 *Transcendence and Spatiality of the Triune Creator*

> The doctrine of the Trinity gradually would be understood to be the exposition of the relations of God *in se*, with scarce reference to God's acts in salvation history. After Augustine, in the period of scholasticism, the eternal, ontological relationships among Father, Son and Holy Spirit would be viewed largely independently of the Incarnation and sending of the Spirit. The divine processions – begetting of the Son, proceeding of the Spirit – would be understood as absolutely interior to God and explicated without reference to any reality 'outside' God.[31]

Peters discusses "Trinity talk in the last half of the twentieth century" as a process of recovering this lost divine relationality. The contention that God exists in the intra-trinitarian relations is not new at all. By the term "perichoresis," the Orthodox theologians since Gregory Nazianzen have sought to express the intra-trinitarian relations in God. But the theological discussion of the last century has extended these intra-trinitarian relations of God to include consideration of his positive relation to the world.

Peters takes Barth as a starting point of his discussion. He quotes Barth's saying that "Without ceasing to be God, He has made Himself a worldly, human, temporal God in relation to this work of His."[32] For Barth, according to Peters, "the Word of God in revelation is not just a word about God but rather is God in the Godself." After discussing Barth, Peters examines Jüngel's principle of correspondence, describing the dilemma which Jüngel tries to overcome as follows: "If, on the one hand, we insist that God is independent – that is, that God is *a se* – and not in any way dependent upon the creation, then it seems that we must deny that relationality is essential to the divine being. On the other hand, if we insist that God's relationship to the creation is constitutive of God's own being, then we make God dependent upon something outside God and hence lose divine aseity."[33] The solution to the dilemma that Jüngel proposes, Peters contends, is to affirm that relationality already exists within the divine being. Relationality already exists within the

31   LaCugna, *God for Us*, p. 81.
32   Peters, *GOD as Trinity*, p. 90. This citation comes from Barth, *Church Dogmatics*, III/2, p. 457. In order to explain this saying of Barth, Peters quotes Colin Gunton in an endnote: "'There is a kind of temporality in God, which preserves both the ontological distinction of God from the world and the real relation God has with it. God's eternity is not non-temporality, but the eternity of the triune life' 'Barth, the Trinity, and Human Freedom,' *Theology Today* 18/1 (April 1986), p. 318" (ibid., p. 212, note 19).
33   Peters, *GOD as Trinity*, p. 91.

The Presence of the Triune God
179

immanent Trinity. According to Peters, "Jüngel's thesis is that God's relationality to the temporal world *corresponds* to the relationality that already exists within the eternal divine life."[34] However, once we have established unity between the immanent Trinity and the economic Trinity, we no longer need Jüngel's correspondence principle.

By what has been called "Rahner's Rule," Rahner has opened the door to the possible identity of God in Godself and God in relationship to others. For Rahner, Peters writes, "The threefold manner by which God relates to us is not 'merely a copy or an analogy' of God's internal threefold relatedness. Rather, it is that relationship proper."[35] On this point, Peters states, "Rahner seems to be going a step beyond Jüngel, for whom there was only a correspondence between the two. Rahner is on the brink of saying that God relates to the Godself through relating to us in the economy of salvation."[36] In relation to Barth, Peters comments on Rahner's thought:

> By identifying the immanent with the economic relations, Rahner opens Barth's door a bit wider so that we might consider how the history of the incarnation as history becomes internal to the divine perichoresis itself. And along with the incarnate Son comes the world that he was destined to save, so that the whole of temporal creation enters into the eternity of God's self-relatedness.[37]

Peters dubs Moltmann's Trinity "Open Trinity."[38] Citing Moltmann's saying that "God suffers with us – God suffers from us – God suffers for us: it is this experience of God that reveals the triune God," Peters explains that "in the surrender to suffering for the sake of sinful humanity, Jesus and the Father experience a new unity with one another in the Spirit. It is an internal unity that is achieved through external historical involvement. History is swept into the divine life because the Trinity is an open Trinity."[39] Peters states that, "What we get with Jürgen

---

34 Peters, *GOD as Trinity*, p. 92.
35 Peters, *GOD as Trinity*, p. 97.
36 Peters, *GOD as Trinity*, p. 97.
37 Peters, *GOD as Trinity*, p. 103.
38 Cf. J. Moltmann, *The Trinity and the Kingdom*, p. 96, contends that "the triunity is open in such a way that the whole creation can be united with it and can be one within it. The union of the divine Trinity is open for the uniting of the whole creation with itself and in itself."
39 Peters, *GOD as Trinity*, p. 103.

180   *Transcendence and Spatiality of the Triune Creator*

Moltmann is perhaps the biggest step yet away from the substantialist unity of God toward a relational unity in which the divine threeness is given priority."[40] According to Peters, "Moltmann rules out a prior ontological unity in the sense that the three hypostases would share the same substance. It is rather an integrative unity, a unification through dynamic mutuality and relationality."[41] Moltmann's doctrine begins with the plurality of the Trinity and only then asks about the unity. It is a "social doctrine of the Trinity."[42] Peters asserts that Moltmann takes the history of God seriously, for, in Moltmann, the history of God is of decisive and enduring value. "Echoing Hegel," Peters says, Moltmann's doctrine of the Trinity "achieves its integrative unity principally by uniting itself with the history of the world [...] What the Moltmann theology should lead us to repudiate is that there exists some sort of second God, a Trinitarian double, a ghostly immanence hovering behind while unaffected by the actual course of divine historical events."[43]

Concerning Leonardo Boff's liberation theology and Whiteheadian process theology, Peters conceives the former as "having run out of gas" and the latter as "a detour away from the main direction of trinitarian discussion" respectively. "Although Boff starts down the road following Moltmann," to Peters's regret, "he quickly decides he cannot travel that far." Thus, Peters comments on Boff's concept of the divine society as follows: "Although Boff wants to work with a correlation between a divine society and a human society on a nonhierarchical basis, the divine society of which he speaks is in fact a monarchy; and because this monarchy is shrouded in eternal mystery apart from the time in which we live, no genuine correlation with human society can be made."[44]

Process theology constructs a relational theory of God. In this sense, according to Peters, it would be in a position to make a substantive contribution to trinitarian thought. However, "[t]he Trinity is clearly not intrinsic to process metaphysics," he states, "so the value of the threefold

---

40   Peters, *GOD as Trinity*, p. 103.
41   Peters, *GOD as Trinity*, p. 104.
42   Moltmann, *The Trinity and the Kingdom*, p. 19, writes, "In distinction to the trinity of substance and to the trinity of subject we shall be attempting to develop a social doctrine of the Trinity."
43   Peters, *GOD as Trinity*, p. 110.
44   Peters, *GOD as Trinity*, p. 114.

The Presence of the Triune God

181

revelation become doubtful."[45] Peters claims that process theologians simply throw in the towel by admitting that "process theology is not interested in formulating distinctions within God for the sake of conforming with traditional trinitarian notions."[46] Process theologians are more interested in the dipolar view of God than the doctrine of the Trinity.

Peters refers to two merits of LaCugna's "God for Us": "The first is LaCugna's insistence that the economy of salvation belongs intrinsically to our understanding of God as Trinity. The second is her understanding of the ecstatic character of the divine movement on the creation and incarnation and consummation."[47] Peters explains the insightful scholarship of LaCugna in relation to Barth: "She knowingly or unknowingly follows the Barthian trail of analysis by arguing that the pre-Nicene biblical apprehension of God's revelation comes already in the form of Father, Son, and Holy Spirit."[48] However, in relation to Rahner, "[w]ith full awareness," Peters writes, "she follows and then passes Rahner when he stops with his rule: the economic Trinity is the immanent Trinity and vice versa."[49] Thus, Peters holds that, "Her own corollary – *theologia* is *oikonomia* and vice versa – leads her to the grand vision of the divine life spanning and incorporating the whole of creation within itself. What we thought was external to God is now internal. The God we thought was *in se*, for the Godself alone, is now God in relation to us."[50] For Peters, LaCugna's dominant and pervasive theme is the relationality of God. But he criticizes her work for lacking a coherent view of the temporality of God. LaCugna leaves underdeveloped, he asserts, "the impact that God's relationship to a world that is temporal might have on God's eternity."[51] According to Peters, this challenge to develop the divine temporality is taken up more directly by Jenson and Pannenberg.

Robert Jenson subscribes to Rahner's rule, as Peters observes, "because it accounts for trinitarian identity and definition determined by the

---

45 Peters, *GOD as Trinity*, p. 115.
46 John B. Cobb, Jr., and David Ray Griffin, *Process Theology: An Introductory Exposition* (Louisville: Westminster/John Knox Press, 1976), p. 110.
47 Peters, *GOD as Trinity*, p. 73.
48 Peters, *GOD as Trinity*, p. 127.
49 Peters, *GOD as Trinity*, p. 127.
50 Peters, *GOD as Trinity*, p. 127.
51 Peters, *GOD as Trinity*, p. 143.

182   *Transcendence and Spatiality of the Triune Creator*

course of events in salvation history."[52] In addition to Rahner's rule, however, Jenson contends for a second rule: "the legitimate theological reason for distinguishing the immanent from the economic Trinity is the freedom of God. It must be the case that God in himself could have been the same God he is, and so triune, had there never been a creation or any saving history of God within creation."[53] Jenson perceives a dilemma between two rules: "Are these two compatible?" Concerning this dilemma, Jenson answers, "they are compatible if we think of the identity of the economic and immanent Trinity as eschatological – that is, if the immanent Trinity is simply the eschatological reality of the economic."[54] According to Peters, "Jenson locates the personhood of God in the totality of God's self-constituting relations with the history of the world. He recognizes that the modern understanding denies that a person can be a monad devoid of relationships. The internal dynamics of a person are intrinsically communal, inseparable from relations with other selves. What we need to grasp is the community of God's personhood and our personhood."[55] Thus, Peters claims, "Jenson seems to be siding with Barth, Jüngel, and Rahner over against Moltmann."[56]

Pannenberg claims that the divinity of each of the three persons of the Trinity is a dependent divinity. Thus, Peters dubs Pannenberg's doctrine of the Trinity a "Dependent Divinity": "The Son is fully himself in relation to the Father; the Father is fully himself in relation to the Son; and the Spirit is fully himself in witness to both."[57] Peters states that, "Pannenberg believes that the reciprocity in the relationship of the divine persons makes room for the constitutive significance of the central events of salvation history for the Godhead of God and thus for time and change within the divine eternity [...] Without this kingdom, God could

---

52   Peters, *GOD as Trinity*, p. 133.
53   Peters, *GOD as Trinity*, p. 133.
54   Peters, *GOD as Trinity*, p. 133.
55   Peters, *GOD as Trinity*, p. 134.
56   Peters, *GOD as Trinity*, p. 134.
57   Peters, *GOD as Trinity*, p. 139. Elsewhere he states, "Tipping closer to Athanasius than to the Cappadocians now, Pannenberg stresses that God is not personal except in one or another of the three persons. When God confronts the world through personal relationship, it will be as Father, as Son, or as Spirit. It will not be in the form of an abstract unity. *God is personal only through one or another of the three hypostases, not as a single ineffable entity*" (ibid., p. 138).

The Presence of the Triune God

183

not be God."[58] Pannenberg identifies God with God's rule. Thus, quoting Pannenberg's saying that "God, through the creation of the world, made himself dependent on this creation and on its history," Peters asserts, "For God to be identified with the divine rule, there needs to be a world that is ruled; and in this sense, God has chosen to become dependent upon his creation."[59] Finally, according to Peters, "Pannenberg argues that it will not be until the eschatological consummation of the world – and then with retroactive power – that the existence and hence the identity of God will be conclusively decided."[60] Peters regards Pannenberg's as the most radical of the post-Barthian relational proposals. By the arguments of the above mentioned theologians, Peters reaches the following conclusion: "God is in the process of self-relating through relating to the world he loves and redeems. God is in the process of constituting himself as a God who is in relationship with what is other than God."[61] Is this conclusion a viable position for Christian theology? To answer this question, I will examine and critically evaluate Pannenberg's relational theology in more detail.

2.3 The Relation of God to the Creation

Concerning the traditional view that "the Father alone is without origin *(anarchos)* among the three persons of the Trinity," or "he is the origin and fount of deity for the Son and Spirit," Pannenberg avers, "this view seems to rule out genuine mutuality in the relations of the trinitarian persons, since it has the order of origin running irreversibly from Father to Son and Spirit."[62] For Pannenberg, "the deity of God is his rule."[63]

---

58 Peters, *GOD as Trinity*, p. 135.
59 Peters, *GOD as Trinity*, p. 140. Panneberg's saying is cited from "Problems of a Trinitarian Doctrine of God," *Dialogue* 26/4 (Fall 1987), p. 255. We find a similar saying in Pannenberg, *Systematic Theology* 1, trans. G. Bromiley (Grand Rapids: Eerdmans, 1991): "Even in his deity, by the creation of the world and the sending of his Son and Spirit to work in it, he has made himself dependent upon the course of history" (p. 329).
60 Peters, *GOD as Trinity*, p. 140.
61 Peters, *GOD as Trinity*, p. 145.
62 Pannenberg, *Systematic Theology* 1, p. 311 f.
63 Pannenberg, *Theology and the Kingdom of God* (Philadelphia: The Westminster Press, 1969), p. 55. As a complement to Rahner's Rule, Roger Olson dubs this say-

184 *Transcendence and Spatiality of the Triune Creator*

Therefore, because God's deity apart from his lordship over the world is impossible in this actual world, he would not be God without ruling this world. Pannenberg thus asserts that God would not be God without this world. God is ontologically related to the world in his nature. If the deity of the Father is dependent upon the Son, the reverse is also true. The deity of the Son is dependent upon the Father. The deity of the one divine person is dependent upon the other two persons.

For Pannenberg, therefore, the deity of the Father is not the source of the Son's deity nor the Spirit's: "The Cappadocians with their thesis that the Father is the fount of deity sometimes come close to a view which threatens the equal deity because they do not expressly add that the Father is the principle of deity only from the perspective of the Son."[64] Nevertheless, contrary to Moltmann, Pannenberg does not give up the monarchy of the Father. As he argues: "In his monarchy the Father is the one God... almost without exception the word 'God' means the Father and not the triune God... the Father is known as the one God by the Son in the Holy Spirit."[65] In addition, "By their work," Pannenberg says, "the Son and Spirit serve the monarchy of the Father. Yet the Father does not have his kingdom or monarchy without the Son and Spirit, but only through them."[66] Through this explanation, Pannenberg tries to confirm the transcendence of the Father.

> Extending the thought of Rahner, one might thus say that creation is brought into the relations of the trinitarian persons and participates in them. Nevertheless, only the persons of the Son and Spirit act directly in creation. The Father acts in the world only through the Son and Spirit. He himself remains transcendent.[67]

It seems to me, however, that this problem is solved by the question of in what sense the divinity of the Father is affected by the historical events of the Son.

Pannenberg considers Jüngel and Moltmann the pioneers of the view that the deity of the Father depends upon the course of events in the

---

ing "Pannenberg's Principle" [Olson, "Wolfhart Panneberg's Doctrine of the Trinity," *Scottish Journal of Theology* 43/2 (1990), pp. 175–206].

64 Pannenberg, *Systematic Theology* 1, p. 322 f.
65 Pannenberg, *Systematic Theology* 1, p. 326.
66 Pannenberg, *Systematic Theology* 1, p. 324.
67 Pannenberg, *Systematic Theology* 1, p. 328.

The Presence of the Triune God

world of creation. They worked out this view with reference to the crucifixion of Jesus. Pannenberg states,

> Through the Son and Spirit [...] the Father [...] stands in relation to the history of the economy of salvation. Even in his deity, by the creation of the world and the sending of his Son and Spirit to work in it, he has made himself dependent upon the course of history [...] The dependence of the deity of the Father upon the course of events in the world of creation was first worked out by Jüngel and then by Moltmann, who illustrated it by the crucifixion of Jesus.[68]

From his view of the crucifixion of the Son, Pannenberg infers that God has made himself dependent upon the history of the world. He also wants to reaffirm the mutual interdependence of the three persons of the Trinity.

> These descriptions of the crucifixion which depict the deity of the Father as affected and questioned by the death of the Son imply that in their intratrinitarian relations the persons depend on one another in respect of their deity as well as their personal being, and that this mutual interdependence affects not only the relations of the Son and Spirit to the Father but also those of the Father to them.[69]

For Pannenberg, the resurrection of Jesus is the key to understanding the eschatological characteristic of the divine revelation. It is the prolepsis of the eschaton. It is also a historical event. The resurrection of Jesus as a historical event provides meaning with the whole process of history. Pannenberg contends that the resurrection of Jesus is not an isolated event but is related to the other historical events of Jesus, especially, the incarnation and the cross:

> the Easter faith of Christians is linked for all time to the earthly history of Jesus of Nazareth, who was rejected by his adversaries as a deceiver and handed over to the Romans for execution, who was raised from the dead by God, and who was thus instituted as Messiah (Acts 2:23f., 36; cf. Rom. 1:4). The resurrection of Jesus is

---

68   Pannenberg, *Systematic Theology* 1, p. 329.
69   Pannenberg, *Systematic Theology* 1, p. 329. Concerning this mutual reciprocity of the intratrinitarian relation, he states as follows: "In so doing I was renewing and developing a view of the reciprocity of trinitarian relations which Athanasius pioneered and Augustine rejected. Materially I was moving in the same direction as Jüngel and Moltmann" (ibid., p. 329, n. 205).

186   *Transcendence and Spatiality of the Triune Creator*

> the basis of Christian faith, yet not as an isolated event, but in its reference back to
> the earthly sending of Jesus and his death on the cross.[70]

As the incarnation and the cross of Jesus not only impacts the Son of God but also the whole Trinity, so also the resurrection has implications for the whole Trinity. Through the resurrection, not only the deity of the Son but also the deity of the Father is confirmed. Thus, for Pannenberg, "the Godhood of the Father depends upon the success of the Son."[71] As Cornelius Buller comments,

> apart from the realization of God's perfect rule in the world, which is mediated
> through the incarnation, death, and resurrection of the Son in the person of Jesus of
> Nazareth, God cannot truly be named Creator. God has determined that the divine
> rule *(Basileia)* be realized in the world through the Son.[72]

According to Pannenberg, the central message that we find from the history of Jesus is the love of God for the world. He cites John 3:16 as well as Pauline passages (Rom. 5:5ff; cf. 8:31-39) as summary statements of the meaning of Jesus in history. The essential content of this history is its expression of God's love. Jesus constantly extended this love in his ministry and continues to do so through his resurrected self. Pannenberg sees God's love as the sum of all other attributes of God.[73] All the personal attributes of God are then various aspects of the love of God: "these attributes – goodness, grace, mercy, righteousness, faithfulnes[s], patience, and wisdom – are all to be seen as aspects of the comprehensive statement that God is love."[74] Pannenberg uses this depiction of God's love to defend the thesis that God has existence in relation to the world and its history: "Because God is love, having once created a world in his freedom, he finally does not have his own existence without this world, but over against it and in it in the process of its ongoing consummation."[75] In his early phase, Pannenberg did not dare contend that the world, as the object of God's lordship, is necessary to God's essential

---

70 Pannenberg, *Systematic Theology* 2, p. 344.
71 Cornelius A. Buller, *The Unity of Nature and History in Pannenberg's Theology* (Lanham, ML: Littlefield Adams Books, 1996), p. 125.
72 Buller, p. 125.
73 Pannenberg, *Systematic Theology* 1, p. 422.
74 Pannenberg, *Systematic Theology* 1, p. 441.
75 Pannenberg, *Systematic Theology* 1, p. 447.

The Presence of the Triune God

187

deity, but finally he did assert that God does not have actual existence without this world.

Pannenberg thinks that his doctrine of the Trinity succeeds in preserving both the transcendence and immanence of God. He thus asserts that, "Only the doctrine of the Trinity permits us so to unite God's transcendence as Father and his immanence in and with his creatures through Son and Spirit that the permanent distinction between God and creature is upheld."[76] However, Pannenberg's doctrine of the Trinity does not hold the independence of God from the history of this world. In Pannenberg's trinitarian scheme, God seems to have himself dependent upon the history. The divinity of the transcendent Father depends upon the activities of the immanent Son and Spirit in the process of history. Thus God can not be God apart from this world. Pannenberg's dependent Trinity thus belies a tendency toward a panentheism.[77]

Although Pannenberg sets himself apart from process theologians in his definition of the creation, he agrees with them in that God respects the independence of creatures. Pannenberg writes:

> Behind the biblical statements, however, there is always the fact that the creatures owe all that they are to God's almighty creative action. Once having called them

---

76 Pannenberg, *Systematic Theology* 1, p. 445.
77 Concerning Pannenberg's version of panentheism, see R. Olson, "Trinity and Eschatology: The Historical Being of God in the Theology of Wolfhart Pannenberg" (Rice University 1984 Ph. D. Dissertation) and my thesis "The Voluntary Panentheism of Wolfhart Pannenberg" (Calvin Theological Seminary 1999 Th. M. Thesis). According to Roger Olson, Pannenberg attempts to preserve the transcendence of God in two ways: (1) by arguing that the self-realization of man, his achievement of his true determination in unity with God, can only come about through active self-differentiation from God. (2) by referring to God as 'future' [R. Olson, "The Human Self-Realization of God: Hegelian Elements in Pannenberg's Christology," *Perspectives in Religious Studies* 13 (1986), p. 219f]. However, it is still questionable whether these efforts of Pannenberg do preserve the transcendence of God effectively. Thus Olson contends: "In distinction from Hegel, he sees the 'ideal' unity of God and man revealed in Jesus Christ as thoroughly eschatological. [I]t is not a timeless reality, but a future state of affairs (the 'Kingdom of God') proleptically realized in Jesus... he [Pannenberg] avoids the pantheistic implications of Hegel's ontology by identifying God with the absolute future of all finite realities. The question that remains is whether Pannenberg can distinguish God from the unity of all finite realities in the future. If not, then the question whether his theology really preserves the transcendence of God remains debatable" (ibid., p. 220).

188   *Transcendence and Spatiality of the Triune Creator*

into existence, the biblical God then respects their independence in a way that is analogous to Whitehead's description. There is truth in the contention that to attain his ends in creation, and especially the end of the creature's own fulfillment, God works by persuasion and not by force. But the patience and humble love with which God seeks his creatures are divine in the sense that they do not proceed from weakness. They are an expression of the love of the Creator, who willed that his creatures should be free and independent.[78]

Pannenberg seems to claim that there might have been no world at all.[79] Yet, the triune God had freely willed to be the creator of this world and then actually created this world. After the creation God then chose to respect the relative independence of his creatures. Therefore, whereas he argues for the contingency of the creation in the origin of this world, Pannenberg seems to deny the absolute contingency of the existence of creatures after creation. He also connects the contingency of this world to the biblical God as the God of history: "The contingency of events and of all created existence is intimately related to the fact that the biblical God is the God of history."[80] Thus, Pannenberg does not differentiate

---

78   Pannenberg, *Systematic Theology* 2 (Grand Rapids: Eerdmans, 1994), p. 16.

79   Pannenberg, *Systematic Theology* 3 (Grand Rapids: Eerdmans, 1998), p. 540, "God does not need a world in order to be himself real, but if a world of finite phenomena exists, then we cannot think of the deity of the one God apart from his rule over the world as his creation." For Pannenberg, however, Ted Peters aptly states that "For God to be identified with the divine rule, there needs to be a world that is ruled" (T. Peters, *GOD as Trinity*, p. 140). Stanley J. Grenz contends that Pannenberg is not a panentheist, since he does avoid "the Hegelian problem of the necessity of the world for the self-realization of God, which is likewise present in certain lines of process theology" [S. J. Grenz, *Reason for Hope: The Systematic Theology of Wolfhart Pannenberg* (Oxford University Press, 1990), p. 138]. For Pannenberg, God could have existed without this world. The creation of this world was not a necessary event. However, this contention of Grenz is due to his misunderstanding of Pannenberg's doctrine of creation. Grenz rightly maintains that for Pannenberg "creation itself is a nonnecessary event" (ibid., p. 139). The creation of this world is a free act of God, contends Pannenberg. But this is only one side of Pannenberg's position. On the one hand this world is a product of God's free will. On the other hand this world is a necessary product of divine nature, as will be shown. If God in himself can be creator, He must create the world in order to actualize that possibility. Therefore, God must create to actualize himself fully. So it is questionable whether Pannenberg avoids "the Hegelian problem of the necessity of the world for the self-realization of God" as Grenz contends.

80   Pannenberg, "Theology and Science," p. 303.

The Presence of the Triune God

"God as the necessary being" from the contingency of the existence of this world and events of the history of this world.

In relation to God's purpose in creation, Pannenberg states that, "The goal of the ways of God is not beyond creation. His acts in the reconciliation and eschatological consummation of the world are oriented to nothing other than the fulfilling of his purpose in creation. But why, then, did he not give creation already its definitive, eschatologically perfected form?"[81] In responding to this question, Pannenberg presents the following: "By ordaining his creation for independence, God took a risk himself, the risk that the autonomy of his creatures would make him seem to be nonessential and even nonexistent [...] by his reconciling action God stands by his creation, and does so indeed in a way that respects his creatures' independence."[82]

Peters correctly indicates the difficulty of Pannenberg's Dependent Divinity. He comments: "The picture one gets here is of a God who jeopardizes his own divinity in order to engage in historical intercourse with created reality. Creation constitutes a divine risk of considerable consequence. Why would a free God choose this destiny? God chooses this path of creative dependence out of love for the world."[83] In a more

---

81 Pannenberg, *Systematic Theology* 3, p. 642.
82 Pannenberg, *Systematic Theology* 3, p. 643. According to Pannenberg, this independence of creatures does not end even in the eschatological consummation. For this is "the condition of the mutuality of eschatological glorification in which creatures are not only glorified but also for their part glorify Jesus and the Father." Thus, he contends that, "we can speak of this mutuality only because creatures have an existence with its own center and characterized by spontaneity in relation to God and their fellows" (ibid.). Hence the glorification of creatures in the system of Pannenberg does not imply their absorption into the life of God: "The independence of the creature is still preserved in its eschatological consummation before God." Hence, he writes, "the creature is not swallowed up by the presence of God" (ibid., p. 555). For Pannenberg, this creaturely independence "is not possible without temporality as a form of existence." Thus, he contends that, "without inclusion of the difference between time and eternity we would have no conception of the process of fashioning an independent and finite being that has its own center" (ibid., p. 643f). In his eschatological vision, Pannenberg still includes the element of temporality. However, "because of the sin that goes along with our being in time, the sin of separation from God, and of the antagonism of creatures among themselves," he states, "the participation of creatures in the eternity of God is possible [...] only on the condition of a radical change" (ibid., p. 607).
83 Peters, *GOD as Trinity*, p. 141.

190   *Transcendence and Spatiality of the Triune Creator*

positive evaluation, Stanley Grenz asserts that Pannenberg "is able to develop a trinitarian understanding of the nature of God in which the triune God is truly involved with the historical process – even to the point of suffering with creation – but is not limited by this process."[84] I would tend more toward Peters' critical position on Pannenberg, agreeing that Pannenberg's model assigns too much freedom to creatures. I agree with the following saying of R. Olson: "Pannenberg's theology has a much stronger affinity to classical theism than to process theology while his clear intention is to overcome that duality in a synthesis."[85] Although he objects to process theology on some points, Pannenberg still wants to preserve its merits, especially the intimate mutual dependence between God and world. In Pannenberg's thought, God in some sense needs the world. God has determined to depend upon the process of the world. Even though Pannenberg's view may be closer to classical theism than to other forms of panentheism, his system is far enough from classical theism so as to slide into the kind of panentheism: "voluntary panentheism."[86]

Admitting that Rahner's principle is valid, LaCugna indicates that there remain fundamental problems of interpretation on the ground of ontological and epistemological issues. Thus, she proposes a question: "Is there a way to preserve a distinction of reason between economic and immanent Trinity without allowing it to devolve into an ontological distinction? This is crucial because if the distinction is ontological, then *theologia* is separated from *oikonomia*. If the distinction is ontological, then *oikonomia* is our means of access to *theologia*, and it is truly *theologia* that is given in *oikonomia*."[87] LaCugna states reservations about Rahner's axiom expressed by theologians such as Piet Schoonenberg, Yves Congar, and Walter Kasper. Schoonenberg explicates some of the consequences of Rahner's axiom in an important set of "theses on the

---

84   Grenz, *Reason for Hope*, p. 72.

85   Olson, "Wolfhart Pannenberg's Doctrine of the Trinity," p. 206.

86   Olson presents Pannenberg's version as eschatological panentheism. See Olson, "Trinity and Eschatology: The Historical Being of God in the Theology of Wolfhart Pannenberg" (Rice University 1984 Ph. D. Dissertation). I think Moltmann is also a voluntary panentheist in that they both emphasize the voluntary freedom of God rather than the necessity of his being as in process theologies. In this sense, Pannenberg and Moltmann are not ontological panentheists like process theologians but voluntary panentheists.

87   LaCugna, *God for Us*, p. 217.

The Presence of the Triune God

trinitarian God." According to LaCugna, Schoonenberg agrees with Rahner that *economia* is the starting point of access to the mystery of *theologia*. The economy is also the means of access to theology. However, "Schoonenberg's theses point to an inherent asymmetry between economic and immanent Trinity; there can be no strict ontological identity because we must leave room for the freedom of divine self-expression in salvation history, *and* the freedom of the recipient to accept the divine self-communication."[88] LaCugna claims that, "Yves Congar advances a similar caution about the 'vice versa' in Rahner's principle. While the economic Trinity is the immanent Trinity, the reverse affirmation requires care. What God is remains ineffable, and not fully identical with God's economic self-expression."[89]

Walter Kasper warns us against two opposite misunderstandings of Rahner's identification. The first one is to understand the economic as a mere temporal manifestation of an eternal immanent Trinity: "History counts. The second person incarnate in Jesus of Nazareth means God exists in the world in a new way. The incarnation implies that God really does become."[90] The second is the dissolution of the immanent Trinity in the economic, as though the eternal Trinity first came into existence in and through history. LaCugna maintains that, "Kasper, Schoonenberg, and Congar, like Rahner, mean to emphasize the *historicity*, which is to say, the genuine *incarnateness*, of God. God's self-expression in the economy of salvation thereby becomes an essential dimension of God's being as God."[91] Kasper reformulates Rahner's rule as follows: "In the economic self-communication the intratrinitarian self-communication is present in the world in a new way, namely, under the veil of historical words, signs, and actions, and ultimately in the figure of the man Jesus of Nazareth."[92] By this reformulation, however, according to the analysis of Ted Peters, "Kasper seems to assume that the internal relations of the immanent Trinity are already fixed in eternity, so that what is new is merely their manifestation under the 'veil' of history; that is, the meaning and significance of the Trinity in history is due solely to its being a manifestation of the Trinity in eternity." Thus Peters contends that,

---

88 LaCugna, *God for Us*, p. 219.
89 LaCugna, *God for Us*, p. 220.
90 Peters, *GOD as Trinity*, p. 102.
91 LaCugna, *God for Us*, p. 220.
92 W. Kasper, *The God of Jesus Christ* (New York: Crossroad, 1986), p. 275f.

192 *Transcendence and Spatiality of the Triune Creator*

"Kasper avoids the second misinterpretation only by retreating to the first misinterpretation."[93]

The above criticism of Peters against the view of Kasper also applies to Pannenberg's thought. Pannenberg denies the view that the trinitarian God is the result of history and achieves reality only with its eschatological consummation. In that sense, Pannenberg rejects "the idea of a divine becoming in history."[94] Since he wants to avoid "the absorption of the immanent Trinity in the economic Trinity," Pannenberg is seeking to provide room for the immanent Trinity. However, this room is provided indirectly through the economic Trinity. Thus, for Pannenberg, the immanent Trinity, that is, the essence of God, has some potentiality. And this immanent Trinity is actualized as the economic Trinity in history. In that sense Pannenberg agrees with Kasper that "to affirm the Trinity of revelation is to affirm the essential Trinity, the trinitarian fellowship of Father, Son, and Spirit from eternity to eternity."[95] For Pannenberg, in some sense, therefore, there is only the economic Trinity.

Unlike Pannenberg, Colin Gunton argues for the importance of the doctrine of the immanent Trinity. He states that, "Because God is, 'before' creation took place, already a being-in-relation, there is no *need* for him to create what is other than himself. He does not need to create, because he is already a *taxis*, order, of loving relations."[96] Gunton acknowledges there might be an objection to this view: "if God does not need the creation in some way or other, he must be a distant and unfeeling monarch." However, Gunton contends that such a contention confuses two points. The first point is "the proper objection to the form that doctrines of aseity have sometimes taken, suggesting as they do the total lack of involvement, of an Epicurean kind, of a completely immutable and unfeeling deity."[97] Yet, "it does not follow that for God to enter into relation with the world he must need it in some way."[98] Concerning the second point, he holds as follows: "far from suggesting an unrelatedness of God to the world, trinitarian theology is based on the belief that God the Father is related to the world through the creating and redeeming action

93 Peters, *GOD as Trinity*, p. 102.
94 Pannenberg, *Systematic Theology* 1, p. 331.
95 Pannenberg, *Systematic Theology* 1, p. 332.
96 Gunton, "Relation and Relativity," p. 97.
97 Gunton, "Relation and Relativity," p. 97.
98 Gunton, "Relation and Relativity," p. 97.

The Presence of the Triune God

of Son and Spirit who are, in Irenaeus' expression, his two hands. The doctrine of the Trinity [...] is indeed *derived* from the involvement of God in creation, reconciliation and redemption."[99] The relation between God and the world is, he avers, "the relation-in-otherness that is conceived with the help of the doctrine of the Trinity, and probably cannot adequately be conceived in other way."[100] Even though "the world is itself, not God, but worldly according to its own measure of being," Gunton states, "it is so by gift of the God who creates and sustains it in such a way that it is itself."[101] This contention leads us to the discussion of the contingency of the world.

Torrance claims that contingence has a double aspect deriving from its theological roots. He explains, "On the one hand, it means that the universe depends entirely upon the grace and wisdom of God for its being and form, and is not a necessary emanation of the divine. On the other hand, it also means that the universe is given a distinctive reality of its own completely differentiated from the self-sufficient reality of God."[102] Thus, Torrance avers, "in virtue of its contingence the universe has an orientation at once toward God and away from him."[103] While theological science is more concerned with the dependence on God of the universe, natural science is more concerned with its independence of God.[104] By contingence, Torrance expresses this dual aspect of contingence, "that as created out of nothing the universe has no self-subsistence and no ultimate stability of its own, but that it is nevertheless endowed with an authentic reality and integrity of its own which must be respected."[105] Unlike definitely process theologians and probably Pannenberg, Torrance argues for "an asymmetrical relation between God and the world, characterized by perfect freedom on God's part and sheer dependence on the world's part."[106] He holds, "The world needs God to be what it is, but God does not need the world to be what he is, the eternally self-existent God who is not dependent on anything other than him-

---

99   Gunton, "Relation and Relativity," p. 97.
100 Gunton, "Relation and Relativity," p. 97.
101 Gunton, "Relation and Relativity," p. 97f.
102 Torrance, *Divine and Contingent Order*, p. 71.
103 Torrance, *Divine and Contingent Order*, p. 71f.
104 Torrance, *Divine and Contingent Order*, p. 72.
105 Torrance, *Divine and Contingent Order*, p. vii.
106 Torrance, *Divine and Contingent Order*, p. 34.

194 *Transcendence and Spatiality of the Triune Creator*

self."[107] Again unlike Pannenberg, Torrance asserts that "to be meaningful contingence must have only a relative and not an absolute independence, for absolute independence would alienate it from any signitive or semantic function."[108] Torrance maintains that "if the alienation of contingent reality from God were wholly actualized, if the universe were to achieve autonomy from its Creator, it would disappear entirely into nothing."[109] This logic finds its full expression in the impassioned atheistic existentialism of Sartre, who would say that "rejection of God is needed to save contingency and freedom."[110] Torrance comments on Sartre's thought:

> Absolute contingence thus replaces in Sartre's view the notion of absolute necessity [...] 'Nothingness haunts being' [...] Thus through this utterly contingent existence of man for himself, cut off from God, there is posited in the very heart of immanence 'a transcendent nothingness' to replace the transcendent God as the ultimate ground of being and its intelligibility.[111]

Thus, Torrance maintains, "the great lesson to be gained from Sartre's atheistic *tour de force* is that the notion of contingence, deprived of any inner ontological intelligibility beyond itself, collapses into disorder and pointlessness."[112]

It is so hard to determine the relationship between God and the world. Whereas the intimate attachment of the world to God results in pantheism or panentheism, the corollary of the total detachment of the world from God is deism. Concerning Pannenberg's position that Peters has reached as a conclusion of his book, I presented two main objections: Pannenberg neglects the importance of the immanent Trinity and gives too much freedom to creatures. Although he is self-sufficient, God is relational not only in himself but also toward this world. The genuine relation between God and the world is possible when the autonomy of creatures is in some sense confirmed. In traditional theism, the world had been intimately attached to God. The universe has been seen to have no autonomy apart from God. Since the Reformation, however, the world

107 Torrance, *Divine and Contingent Order*, p. 34.
108 Torrance, *Divine and Contingent Order*, p. 107.
109 Torrance, *Divine and Contingent Order*, p. 119.
110 Torrance, *Divine and Contingent Order*, p. 105.
111 Torrance, *Divine and Contingent Order*, p. 106.
112 Torrance, *Divine and Contingent Order*, p. 107.

The Presence of the Triune God

began to be detached from God. The universe came to be seen as autonomous. In this scheme, the development of natural sciences was possible. A corollary of this detachment of the world from God was deism. Yet further development of the natural sciences has suggested that this is not the whole story. Deism presupposes so-called Newtonian-ism or Laplacian determinism. This deterministic world view collapsed with new discoveries of natural sciences. Thus God came to be seen once again as having possibility to be present within, interact with, or act in the universe.

3. Agency

In this section, I will discuss the modern theological scholarship related to Dyrness's second category, agency. Narrowing down this category into the divine action, I will provide the second meaning of God's spatiality as follows: God needs space, room, or gaps for his action in this universe. Concerning the importance of the topic of divine action, Nancey Murphy contends: "I predict that although significant progress has been made in recontextualizing theology for the postmodern age, no ultimately satisfying program will appear without a solution to the problem of divine action."[113] Among several solutions to the issue of divine action, I will pursue the views of three theologians: Arthur Peacocke, Nancey Murphy, and John Polkinghorne. On some points, they converge, while on others they are divergent. Learning from their views, I will contend for a general position which is related to God's spatiality. The following saying of Gunton will provide a guide to my discussion of this section: "Developments in such areas as quantum theory, biology and chaos theory are held to signal conceptions of a new openness in the shape of reality which enables theology to conceive not a God of the gaps, but God broadly involved in the formulation of the world."[114] Pea-

---

113 N. Murphy, *Beyond Liberalism and Fundamentalism: How Modern and Postmodern Philosophy Set the Theological Agenda* (Valley Forge, PA: Trinity Press International, 1996), p. 153.

114 C. E. Gunton, "The Doctrine of Creation," C. E. Gunton ed., *The Cambridge Companion to Christian Doctrine* (Cambridge University Press, 1997), p. 154. But I use

196 *Transcendence and Spatiality of the Triune Creator*

cocke was once a biologist. From this frame of reference, he proposes chance as the method of God's action. Murphy presents quantum indeterminacy as the room, or space, of divine action. Polkinghorne proposes chaos theory as consistent with God's action.

Subsection 1 deals with Peacocke's objections to the Laplacian determinism which dominated the modern world-view. Peacocke convincingly demonstrates that neither deism nor interventionism is the appropriate solution to determinism. Peacocke also presents various objections to interventionism, which also presupposes determinism. Despite opposition to determinism, however, Peacocke can leave room for talk of causes. He presents chance as a room for God's action. Accordingly, subsection 2 examines Peacocke's top-down causation (recently expressed as "whole-part constraint"). Peacocke thinks the metaphor of the world as God's body is useful to explain divine action in the world. But the earlier versions of this analogy are based upon a mind-body dualism. Thus he proposes top-down causation/whole-part constraint as his own model of divine action based upon a non-dualistic view. Subsection 3 suggests two alternatives to Peacocke's top-down causation alone. Murphy opts for quantum indeterminacy as one of the scopes for speaking of the divine action. In opposition to Murphy, Polkinghorne argues for chaos theory as a way to account for God's action. While quantum theory pertains to a micro-world phenomenon, chaos theory pertains to macro-world phenomenon.

3.1 Interventionism

Even though he did not intend this outcome, the model of nature which the classical physics of Newton suggested to us was one of the deterministic mechanism.[115] In this mechanistic worldview the concept of God as a clock-maker appeared. Thus deism held the position that God created

---

the term of "gaps" not in the sense of the God of the gaps but in a neutral sense. Cf. Thomas F. Tracy, "Particular Providence and the God of the Gaps," *Chaos and Complexity: Scientific Perspectives on Divine Action* (Vatican City State: Vatican Observatory Publications, 1997, 2nd Ed.).

115 Cf. Edward B. Davis, "Newton's Rejection of the 'Newtonian World View': The Role of Divine Will in Newton's Natural Philosophy," *Fides et Historia* 22 (Sum 1990), pp. 6–20.

The Presence of the Triune God

this world at the beginning but that he has no further causal influence on the world beyond creation. This conclusion was the logical outcome of "the difficulty of conceiving incursions ('interventions') of God into the world."[116]

Thus today, having overcome Spinoza's pantheism, the refutation of deism is still one of the most important tasks for the scientist-theologians. As Philip Clayton asserts: "For the scientist-theologians writing in this area, the most pressing issue is how to avoid de facto deism – not merely by calling it unorthodox and expressing their dislike of it, but by eventually showing why it is an unnecessary conclusion to draw from contemporary science."[117] Along these lines, Arthur Peacocke also submits one of the most cogent refutations of deism: "the affirmations that God is continuously acting creatively in the world through its natural processes, that he acts to redeem and save in history and that he shapes the course of individual lives are all central to Christian belief."[118]

In Peacocke's defense of theism, he emphasizes that he does not need to adopt an "interventionist" framework. Thus, he would reject a position like that of David Brown, who mistakenly assumes that theism "necessarily involves God 'intervening' in his created world as a kind of *deus ex machina* interrupting otherwise orderly created processes."[119] Instead of "intervention," Peacocke prefers the term "interaction" to describe God's involvement with the world:

> As I hope will become clear, it is possible to think of this continual, and effective, interaction of God with the world in ways that are not crudely 'interventionist' in the manner I have just depicted, yet can still serve to underpin the doctrines Brown is rightly, in my view, concerned to defend.[120]

For interventionists, the omnipotence of God implies that God "has the power to achieve particular purposes, to 'set aside' his own laws opera-

---

116 Philip Clayton, *God and Contemporary Science* (Grand Rapids: Eerdmans, 1997), p. 188.
117 Clayton, p. 192.
118 Arthur Peacocke, *Theology for a Scientific Age: Being and Becoming – Natural, Divine, and Human* (Minneapolis: Fortress Press, 1993, Enlarged Edition), p. 137. Hereafter *"TSA."*
119 Peacocke, *TSA*, p. 140.
120 Peacocke, *TSA*, p. 140.

198   *Transcendence and Spatiality of the Triune Creator*

tive in the world he has created."[121] Peacocke, as well as Ian Barbour, John Polkinghorne, Nancey Murphy, and William Alston, however, are all uncomfortable with this interventionist view.[122]

The first reason that Peacocke presents for our rejection of interventionism, is that ironically interventionism shares the same mechanistic view of the world as deism:

> The idea of an 'intervening' God appears to presuppose that God is in some sense 'outside' the created world and has in some way, not specified, to come back into it to achieve his purposes. Yet we have already found good reason why the presence and immanence of God within the created order needs to be re-emphasized. A mechanistic view of the world inevitably engenders, as it did in the eighteenth century, a deistic view of God and of his relation to it.[123]

In other words the interventionist view cannot emphasize the immanence of God in this world fully as deism denies it. According to Peacocke, "The belief that God intervenes in the natural nexus for the good or ill of individuals and societies" stems from a Newtonian and determinist view

---

121   Peacocke, *TSA*, p. 141.

122   Cf. J. Polkinghorne, *Scientists as Theologians: A Comparison of the Writings of Ian Barbour, Arthur Peacocke and John Polkinghorne* (London: SPCK, 1996), p. 41, states that, "We all refuse the word 'intervention', and accept the word 'interaction', as the way to speak about divine acts." See also N. Murphy, *Beyond Liberalism and Fundamentalism: How Modern and Postmodern Philosophy Set the Theological Agenda* (Valley Forge, PA: Trinity Press International, 1996). Concerning Alston's position on the interventionist view, see William P. Alston, "God's Action in the World," *Evolution and Creation*, ed., Ernan McMullin (Notre Dame, IN: University of Notre Dame Press, 1985), pp. 197–220. But Alston recently changes his position on the use of the term divine "interventions" or "interferences." See Alston, "Divine Action: Shadow or Substance?" *The God Who Acts: Philosophical and Theological Explorations*, ed., Thomas F. Tracy (University Park, PA: The Pennsylvania State University Press, 1994), p. 45f, "Talk of divine 'intervention' stems from a deist picture of God as 'outside' His creation, making quick forays or incursions from time to time and then retreating to His distant observation post. But despite these qualms, I shall frequently use the term 'divine intervention,' since I know of no comparably concise way of referring to the intentional divine production of a particular state of affairs at a particular place and time."

123   Peacocke, *TSA*, p. 141f.

The Presence of the Triune God

of the universe with "a wholly transcendent God as the great law-giver."[124]

Peacocke further relates the divine interventionist view to the concept of "God of the gaps." He sees the imprudence of this view in the historical patterns of science gradually, filling in the gaps, making God irrelevant, or otiose.[125] Peacocke defends a different view of a God of the *permanent*, and *unclosable*, gaps "'within' the flexibility we find in these [...] unpredictable situations in a way that could never be detected by us."[126] In this view, Peacocke maintains, "there would then be no possibility of such a God being squeezed out by increases in scientific knowledge."[127]

Even though he acknowledges that this mode of divine action would not be inconsistent with any valid scientific knowledge of a given situation, Peacocke indicates the pitfall of such an explanation of God's action: it is not actually different from that of divine intervention. "The only difference in this proposal from that of earlier ones postulating divine intervention would be that," he states, "given our recent recognition of the actual unpredictability, on our part, of many natural systems, God's intervention would always be hidden from us."[128]

Contrary to both deism and interventionism, Peacocke holds that God acts continuously in this world and that he does not intervene in ways which break the laws of nature; rather, these very laws express his divine will for creation:

> God, I am suggesting, is thus to be conceived of as all the time the continuing supra-personal, unifying, unitive Agent acting, often selectively, upon all-that-is, as

---

124 A. Peacocke, "Chance and Law in Irreversible Thermodynamics, Theoretical Biology, and Theology," *Chaos and Complexity: Scientific Perspectives on Divine Action*, eds. R. J. Russell, N. Murphy and A. R. Peacocke (Vatican City State: Vatican Observatory Publications/Berkeley, CA: The Center for Theology and the Natural Sciences, 1997, 2nd ed.), p. 142f.

125 A. Peacocke, "God's Interaction with the World: The Implications of Deterministic 'Chaos' and of Interconnected and Interdependent Complexity," *Chaos and Complexity: Scientific Perspectives on Divine Action*, p. 277.

126 Peacocke, "God's Interaction with the World," p. 277.

127 Peacocke, "God's Interaction with the World," p. 277.

128 Peacocke, "God's Interaction with the World," p. 278.

200 *Transcendence and Spatiality of the Triune Creator*

> God's own self purposes... God, so conceived, does not intervene to break the causal chains that go from "bottom-up," from the micro- to the macro levels.[129]

Peacocke's own position on divine action comes from relatively new scientific understandings of the world. This world, or universe, is not deterministically closed but is flexibly open-ended.

Peacocke presents his definition of "world"/"all-that-is" as follows: "Here, and in what follows, by both 'world' and 'all-that-is'[,] I intend to refer to everything apart from God, both terrestrially and cosmologically. I regard the world/all-that-is, in this context, as an interconnected and interdependent system, but with, of course, great variation in the strength of mutual inter-coupling."[130] Along with other contemporary scientists, he declares that the age of deterministic mechanism is gone, but that this newer vision of world has a degree of openness and flexibility within a lawlike framework. Thus Peacocke states:

> *Then*, as we saw earlier, the natural world was regarded as mechanically determined and predictable from any given state by means of laws of all-embracing scope: *now*, the world is regarded rather as the scene of the interplay of chance and of statistical, as well as causal, uniformity in which there is indeterminacy at the *micro*-level and unpredictability at the *macro*-level, especially that of the biological.[131]

Upon this worldview, therefore, Peacocke contends that "the concept of God as the deterministic Law-Giver prescribing *all* in advance" seems to be no longer adequate and even false.[132]

Peacocke, however, opposes to the contentions of Jacque Monod that "'pure chance, absolutely free but blind' lies 'at the very root of the stupendous edifice of evolution'" and that we cannot have any hope to find "any meaning in the universe, in general, and in the human presence in it, in particular."[133] Peacocke criticizes Monod in this way: "this randomness of molecular event in relation to biological consequence does not have to be raised to the level of a metaphysical principle interpreting the universe."[134] Though now we live in the world as possessing a more

---

129 Peacocke, "God's Interaction with the World," p. 286.
130 Peacocke, "God's Interaction with the World," p. 276, n. 33.
131 Peacocke, *Intimations of Reality*, p. 57.
132 Peacocke, *Intimations of Reality*, p. 65.
133 Peacocke, *Intimations of Reality*, p. 70.
134 Peacocke, *Intimations of Reality*, p. 70.

The Presence of the Triune God

201

open-ended character, we can find "a much looser coupling between any two given events" and "interlocking networks of statistical relationships, both at the subatomic level because of the significance in that domain of the Heisenberg Uncertainty Principle, and in the macroscopic world of biology and the cosmos."[135]

In a section titled "Living with Chance," in his essay, "Chance and Law," Peacocke explains the term "patterned chance" as set forth by Rustum Roy: "He [Roy] urged us to accept that the world displays patterned chance. It is not a chaos, for there is only a 'loose coupling' through statistical laws and patterns, which still allows talk of 'causes.' In human life we must accept... that reality has a dimension of chance interwoven with a dimension of causality – and that through such interweaving we came to be here and new forms of existence can arise."[136] Therefore, we have to note that "the demise of determinism in its strict Laplacian form has not vitiated entirely the concept of causality."[137] In other words, "causality, as such, is *not* eliminated, only classical, deterministic (Laplacian) causality."[138]

In various places, Peacocke suggests the mutual interplay of chance and law (necessity or determinism) as the mode of God's creativity.[139] Peacocke contends, *"God is the ultimate ground and source of both law ('necessity') and 'chance'* [...] God creates in the world *through* what we call 'chance' operating within the created order, each stage of which constitutes the launching pad for the next."[140] More positively, he asserts, "chance is the search radar of God."[141]

In short, for Peacocke, the world in which we now live is not deterministically causal, as viewed by the Laplacian determinism, but nevertheless allows talk of "causes." For chance, which plays a role as the mode of God's creativity, does not mean chaos. It is in this sense that

---

135 Peacocke, *Intimations of Reality*, p. 64f.
136 Peacocke, "Chance and Law," p. 142.
137 Peacocke, *Intimations of Reality*, p. 64.
138 Peacocke, "God's Interaction with the World," p. 266. Concerning "causality in scientific accounts of natural sequences of events," Peacocke contends that it "is only reliably attributable when some underlying relationships of an intelligible kind between the successive forms of the entities have been discovered, over and beyond mere conjunction as such" (ibid., p. 265).
139 Peacocke, "Chance and Law," p. 142 and *Intimations of Reality*, p. 71.
140 Peacocke, *TSA*, p. 119.
141 Peacocke, *Intimations of Reality*, p. 71.

202 *Transcendence and Spatiality of the Triune Creator*

Peacocke can say, this world possesses a degree of openness and flexibility "within a lawlike framework." However, Peacocke concludes:

> This newly-won awareness of the unpredictability, open-endedness, and flexibility inherent in many natural processes and systems does not, of itself, help directly to illuminate the "causal joint" of how God acts in the world, that is, the nature of the interface between God and all-that-is – much as it alters our interpretation of the meaning of what is actually going on in the world.[142]

Thus, we live in the world in which we should refuse determinism but we can still talk of "causes." In this world view, we can allow for a kind of space, room, or gaps for God's action. Yet, it is still difficult to describe the means by which God acts.

3.2 Top-Down Causation

Peacocke asserts that top-down causation in no way derogates from the continued recognition of the effects of its components on the state of the system as a whole (i.e., of "bottom-up" effects).[143] However, because hardly anyone since the rise of reductionist scientific methodologies doubts the significance of bottom-up causation, he states, the need for recognition of top-down causation is greater. For Peacocke, the lack of a proper recognition of top-down causation has unfortunately often inhibited the development of concepts appropriate to the more complex levels of the hierarchy of natural systems.[144]

According to Peacocke, Donald T. Campbell and Roger W. Sperry are the exemplary proponents of this "downward" (or "top-down") causation. Peacocke explains what Campbell calls "downward causation" as follows: "the network of relationships that constitute the temporal evolu-

---

142 Peacocke, "God's Interaction with the World," p. 281f. We find a similar comment in Peacocke, *TSA*, p. 156.
143 See Peacocke, "God's Interaction with the World," p. 272. There are two opinions about causality when it is applied to systems and their parts: "bottom-up" causation and "top-down" (or "downward") causation. The former contends "the effect on the propensities and behavior of the system of the properties and behavior of its constituent units." However, the latter recognizes "an influence of the state of the system as a whole on the behavior of its component units."
144 Peacocke, "God's Interaction with the World," p. 273.

The Presence of the Triune God

tionary development and the behavior pattern of the whole organism is determining what particular DNA sequence is present at the controlling point in its general material in the evolved organism."[145]

Peacocke has suggested the more refined term "whole-part constraint" for the notion of "downward-" or "top-down causation" which he used more generally in *Theology for a Scientific Age*. He explains the honing of this term as follows:

> The "downward/top-down causation" terminology can, I have found, be misleading since it is actually meant to denote an effect of the state of the system as a whole on its constituent parts, such as the constraints on the parts of the boundary conditions of the system as a whole – more broadly, the constraints of actually being in the interacting, cooperative network of that particular, whole system. The word 'causation' is not really appropriate for describing such situations...[146]

So, in his more recent work, Peacocke has presented the whole-part constraint as his model for God's interaction with the world:

> If God interacts with the "world" at a supervenient level of totality, then God, by affecting the state of the world-as-a-whole, could, on the model of whole-part constraint relationships in complex systems, be envisaged as able to exercise constraints upon events in the myriad sub-levels of existence that constitute that "world" without abrogating the laws and regularities that specifically pertain to them – and this without "intervening" within the unpredictabilities we have noted.[147]

As an example for describing the whole-part constraint of one "level" upon another, Peacocke mentions the relation of the brain state as a whole to the individual neurons. In this discussion, he maintains that "combining a non-dualist account of the human person and of the mind-body relation with the idea of whole-part constraint illuminates the way in which states of the brain-as-a-whole could have effects at the level of

---

145 Peacocke, "God's Interaction with the World," p. 274.
146 Peacocke, "God's Interaction with the World," p. 272, n. 22. In this book, however, I use both terms interchangeably.
147 Peacocke, "God's Interaction with the World," p. 283. Compare this with the following parallel saying in *TSA*: "If God interacts with the 'world' at this supervenient level of totality, then he could be causatively effective in a 'top-down' manner without abrogating the laws and regularities (and the unpredictabilities we have noted) that operate at the myriad sub-levels of existence that constitute that 'world'" (p. 159).

204 *Transcendence and Spatiality of the Triune Creator*

neurons, and so of bodily action."[148] Peacocke suggests that this combination finally provides a fruitful model for illuminating how we might think of God's interaction with the world. So he asserts: "This holistic mode of action on and influence in the world is God's alone and distinctive of God. God's interaction with the whole and the constraints God exerts upon it could thereby shape and direct events at lesser levels so that the divine purposes are not ultimately frustrated."[149]

Even though this model may help to conceive of "how God's transcendence and immanence might be held coherently together as a transcendence-in-immanence," Peacocke admits that it has its limitations, or at least an inevitably negative aspect: "For, in a human being, the 'I' does not transcend the body ontologically in the way that God transcends the world and must therefore be an influence in the world-state from 'outside' in the sense of having a distinctively different ontological status."[150]

According to Peacocke, "the notion of God's relation to the world as being at least analogous to the relation of the human mind to the human body (God:world::mind:body)" has a long history in Christian theology and has been extensively used by process theologians in recent years. He argues that this analogy was earlier always based "on dualist assumptions about man" and "on interventionist assumptions about God's action in the world."[151] However, he asserts, the modern combination of holistic conceptions of the human person with this analogy facilitates a non-interventionist way of thinking of God's agency in the world and a more intimate and internal form for God's knowledge of the world.[152]

Peacocke cites Sallie McFague and Grace Jantzen as scholars who use the metaphor of the world as God's body. He evaluates the thought of McFague as world-affirming and ecologically responsible in stressing God's genuine involvement in the suffering and joys of the world, and

---

148 Peacocke, "God's Interaction with the World," p. 284.
149 Peacocke, "God's Interaction with the World," p. 286. See also Peacocke, *TSA*, p. 161. There Peacocke writes, "In this model, God would be regarded as exerting continuously top-down causative influences on the world-as-a-whole in a way analogously to that whereby we in our thinking can exert effects on our bodies in a 'top-down' manner."
150 Peacocke, "God's Interaction with the World," p. 285.
151 Peacocke, *Intimations of Reality*, p. 76.
152 Peacocke, *TSA*, p. 167.

The Presence of the Triune God

Peacocke praises McFague for not losing sight of the metaphorical character of this language.[153] In comparison with McFague, Peacocke maintains, "Grace Jantzen goes further and regards the world as God's body in a much stronger ontological sense." For her, thus, "God is no longer 'incorporeal,' for God is embodied in the world as the medium of God's life and action." Therefore, in the end, Peacocke criticizes, "her position steers too near to a pantheism in which all-that-is becomes identified so closely with God that God's role as Creator is compromised."[154]

In response to the criticism that top-down causation/whole-part constraint has a tendency toward postulating the "God of the deists," Peacocke rejoins by stressing that the creative processes of the natural order, involving chance operating in a lawlike framework, are themselves the immanent creative activity of God. To help us to imagine this, he presents "a musical analogy."[155] According to Peacocke, music is "the art form *par excellence* into which time enters as an inherent, essential and constitutive element" and is "especially appropriate as a source of models for divine creativity for a number of reasons."[156] Especially "the model of the world process as the unfolding music of the divine Creator-Composer," he holds, helps us to imagine a little better "what we might mean by God's immanence in the world." Thus Peacocke avers,

> There is no doubt of the 'transcendence' of the composer in relation to the music he creates – he gives it existence and without the composer it would not be at all [...] The whole experience is one of profound communication from composer to listener. This very closely models, I am suggesting, God's immanence in creation and God's self-communication in and through what he is creating [...] God, to use language usually applied in sacramental theology, is 'in, with and under' all-that-is and all-that-goes-on.[157]

---

153 Peacocke, *TSA*, p. 167f. See also McFague, *Models of God* (London: SCM Press, 1987) and *The Body of God* (Minneapolis: Fortress Press, 1993).

154 Peacocke, *TSA*, p. 168. See also Jantzen, *God's World, God's Body* (London: Darton, Longman and Todd, 1984).

155 Peacocke, *TSA*, p. 173. Peacocke refers to Gunton as a theologian who uses this musical analogy. He states: "The fascinating nature of music has also provided Colin Gunton with concepts useful in interpreting the Christian doctrine of the Incarnation (in *Yesterday and Today: A Study of Continuities in Christology*, Darton, longman and Todd, London, 1983, pp. 115ff)" (Peacocke, *TSA*, p. 376, note 130).

156 Peacocke, *TSA*, p. 173.

157 Peacocke, *TSA*, p. 176.

206   *Transcendence and Spatiality of the Triune Creator*

It also gives rich and fruitful insights "both into the relation of God as Creator to the creation and into our human apprehension of and participation in the creative process."[158]

Top-down causation/whole-part constraint has, Peacocke writes, "the merit of allowing us to understand how initiating divine action on the state of the world-as-a-whole can itself have consequences for individual events and entities within that world."[159] Thus he asserts, "such divine causative, constraining influence would never be observed by us as a divine 'intervention,' that is, as an interference with the course of nature and as a setting aside of its natural, regular relationships."[160]

3.3 Bottom-Up Causation Revisited

While Nancey Murphy does not refer directly to the issue of God's embodiment, she basically supports Peacocke's positions on interventionism and divine action. In her book, *Beyond Liberalism and Fundamentalism*, Murphy presents her own view on divine action, eschewing both interventionism and immanentism. She defines interventionism as the position that "in addition to God's creative activity, which includes ordaining the laws of nature, God occasionally violates or suspends those very laws in order to bring about an extraordinary event."[161] She claims this interventionist view of God's action is one result of scriptural foundationalism. In this connection, Murphy observes, "Scripture itself is a miracle, understood in interventionist terms."[162] Murphy opposes this position as implying that God's revelation must violate the laws of nature.[163] Murphy, furthermore, criticizes the logic underlying the interven-

---

158 Peacocke, *TSA*, p. 177.
159 Peacocke, "God's Interaction with the World," p. 287. See also Peacocke, *TSA*, p. 163: "The notion of top-down causation has the merit of allowing us to understand how initiating divine action on the state of the world-as-a-whole can itself have a causative effect upon individual events and entities within that world."
160 Peacocke, "God's Interaction with the World," p. 287.
161 Murphy, *Beyond Liberalism and Fundamentalism*, p. 68. This book was originally written as a part of her Rockwell Lectures at Rice University in 1994.
162 Murphy, *Beyond Liberalism and Fundamentalism*, p. 77.
163 Philip Clayton contends that we can use this term in a neutral way not requiring the breaking of the laws of nature. See Clayton, *God and Contemporary Science* (Grand Rapids: Eerdmans, 1997), p. 228, note).

The Presence of the Triune God

tionist account of divine action: "it does not seem to neglect the qualitative difference between God's action and ordinary created causes. To make God a force that moves physical objects seems to make God a part of the system of physical forces. And if God is such a force, we ought to be able to measure divine action just as we do the forces of nature."[164]

Murphy observes, with more hope, that "a variety of scholars from all over the theological map are working toward new conceptions of divine action that allow for special divine acts but at the same time recognize the integrity of the created order."[165] At the same time, however, she refutes "immanentism"[166] and its champion, Gordon Kaufman. According to Murphy, "Kaufman claims that particular acts of God performed from time to time in history and nature are not just improbable or difficult to believe, but 'literally inconceivable.'"[167] Maurice Wiles also holds an immanentist view of divine action. Wiles's position, Murphy claims, "is quite similar to Kaufman's in emphasizing that God enacts the whole of history but does not perform any special divine acts."[168] Philip Clayton also comments on Wiles's immanentism by noting its advantages and its drawbacks. This view has the advantage of making God causally ubiquitous in the world. Yet, this view suggests that "the God who does everything is a God who does nothing."[169] For Murphy, the problem with an immanentist view of divine action is that "it either removes the aspect of

---

164 Murphy, *Beyond Liberalism and Fundamentalism*, p. 81.

165 Murphy, *Anglo-American Postmodernity: Philosophical Perspectives on Science, Religion, and Ethics* (Boulder, CO: Westview Press, 1997), p. 118.

166 Cf. David F. Ford, *Modern Theologians: An Introduction to Christian Theology in the Twentieth Century* (Oxford: Blackwell, 1997), p. 743, comments on "immanentism" as follows: "The view that any events or phenomena within the world are explicable in terms of other events or phenomena within the world, thus excluding any direct agency by God or other supernatural powers; or a view of God which stresses his *immanence* or indwelling in the world rather than his *transcendence*."

167 Murphy, *Beyond Liberalism and Fundamentalism*, p. 73.

168 Murphy, *Beyond Liberalism and Fundamentalism*, p. 77.

169 Clayton, *God and Contemporary Science*, p. 196. The direct explanation of Clayton on Wiles's view is that "although God does not carry out any specific actions in the world subsequent to creation, the continued existence of the universe should be understood as a single composite act of God" (ibid.). Concerning another default position, Clayton contends as follows: "the theologian might give up altogether on the claim that God acts in the world. Such a mover appears also to move outside of Christian theism, at least in the sense in which the tradition has understood it, siding instead with deism, pantheism or atheism" (ibid.).

208 *Transcendence and Spatiality of the Triune Creator*

intention from God's act altogether, or else makes every event equally intentional – devastating earthquakes and the Holocaust as well as the growth of crops and the birth of Jesus."[170]

Murphy demonstrates that both interventionism and immanentism are flawed by means of presupposing the foundationalism of modern philosophy.[171] Her alternative view of divine action is founded on the metaphysical holism of postmodern philosophy.[172] Peters describes Murphy's approach as a "holistic or top-down or supervenient approach."[173] Murphy's view is drawn "from a number of philosophers in the later Wittgenstein or Anglo-American school, especially the concept of the web of beliefs in the work of Willard Van Orman Quine."[174] In contrast to the reductionist tendency of modern science, Murphy proposes a "nonreductive physicalism."[175] The crucial meta-physical assumption of reductionism, according to her, is that *"the parts of an entity or system determine the character and behavior of the whole, and not vice versa."*[176] The goal of the reductionist program is to explain the behavior of entities at higher

170 Murphy, *Beyond Liberalism and Fundamentalism*, p. 81.
171 Cf. Ford, *Modern Theologians*, p. 741, explains "foundationalism" as follows: "Thesis that valid philosophical or theological truth-claims are founded on self-evident propositions or on truths of experience (or a combination of the two)."
172 Concerning the differences between modern and postmodern philosophy, see N. Murphy & James Wm. McClendon Jr., "Distinguishing Modern and Postmodern Theologies," *Modern Theology* 5 (April 1989): pp. 191–214 and Murphy, *Anglo-American Postmodernity*, pp. 7–35.
173 *Dialog: A Journal of Theology* 36 (Fall 1997), p. 318.
174 *Dialog: A Journal of Theology* 36, p. 317. Concerning Quine's holist epistemology, W. V. O. Quine, "Two Dogmas of Empiricism," *From a Logical Point of View: Logico-Philosophical Essays* (Cambridge, MA: Harvard University Press, 1953), chap. 2.
175 Regarding the advantage of the term "nonreductive physicalism," Murphy states, "This term has the advantage both of being widely recognized in contemporary philosophy of mind and also, for Christians, of avoiding the atheistic connotations of *naturalism* and *materialism*" (*Beyond Liberalism and Fundamentalism*, p. 136). She writes, "Nonreductive physicalism has been an available option at least since the 1930s. However, it is only now that philosophers are beginning to get clear causation in science should be expected to increase philosophical interest in this area. Apparently only a few theologians are aware of these developments or of their potential for solving the modern problem of divine action. In addition, much needs to be done to appropriate theologically the radical metaphysical changes called for by quantum theory" (ibid., p. 153).
176 Murphy, *Beyond Liberalism and Fundamentalism*, p. 65.

The Presence of the Triune God 209

levels in terms of their parts. In this view, the hierarchy of the sciences is understood like this: "higher sciences study more complex systems composed of the entities studied at the next level downward."[177] She presents a hierarchy of the natural and human sciences, including cosmology, completed by a metaphysical/ theological layer (see figure 1).[178]

Figure 1

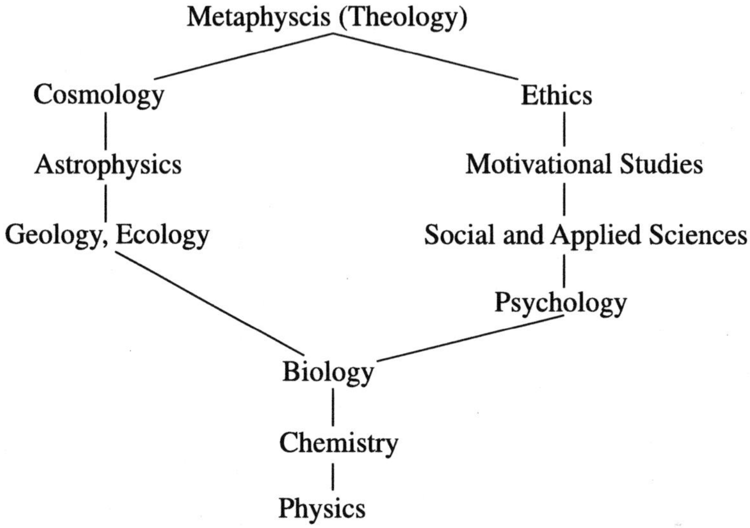

Metaphyscis (Theology)

Cosmology   Ethics

Astrophysics   Motivational Studies

Geology, Ecology   Social and Applied Sciences

                             Psychology

Biology

Chemistry

Physics

However, Murphy contends, "beginning in this century there has been a different way of conceiving this same hierarchy – a nonreductionist view."[179] In this view, the hierarchy of sciences is understood like this: "interactions at the lower levels cannot be predicted by looking at the structure of those levels alone. Higher-level variables, which cannot be reduced to lower properties or processes, have genuine causal impact."[180] In this context, in order to clarify that reductionism in science has limits,

---

177 Murphy, *Beyond Liberalism and Fundamentalism*, p. 136.
178 Murphy, *Reconciling Theology and Science Perspective: A Radical Reformation Perspective* (Kitchener, Ont: Pandora Press1997), p. 80; Murphy and George F. R. Ellis, *On the Moral Nature of the Universe: Theology, Cosmology, and Ethics* (Minneapolis: Fortress Press, 1996), p. 204.
179 Murphy, *Beyond Liberalism and Fundamentalism*, p. 136.
180 Murphy, *Beyond Liberalism and Fundamentalism*, p. 140.

210 *Transcendence and Spatiality of the Triune Creator*

Murphy adopts the term *supervenience*. She defines supervenience as follows: "for any two properties *A* and *B*, where *B* is a higher-level property than *A*, *B* supervenes on *A* if and only if something's being *A* in circumstance *c* constitutes its being *B*."[181] Elsewhere, she explains, "this term refers to a dependence relation between two kinds or levels of properties, such that an entity possesses the high-level property (or properties)."[182] Thus she maintains that "a 'top-down' analysis must be considered in addition to a 'bottom-up' analysis."[183] Murphy refers to Peacocke as both "an effective current proponent of the nonreductive view"[184] and a supporter of "top-down causation within the created order."[185]

According to Murphy, there were two quite different (though related) problems which modern attempts confronted to give an account of divine action: "One was the assumption that God would have to violate or suspend the laws of nature to bring about any *special* divine act [...] The other problem was the question how God could act in a universe where all causes were believed ultimately to be natural forces if God was not a natural force."[186] She thinks that the first problem can be solved by the recovery of top-down causation, because the problem originally was created by the mistaken assumption that all causation was bottom-up. However, Murphy holds that the second problem is not yet solved. She presents the following reason and a suggestion for moving beyond the difficulty: "current science involves conservation of matter and energy," she writes, "and we still have the (analogous) problem of explaining how God's action is to be reconciled with this principle. I believe that the way

---

181 Murphy, *Anglo-American Postmodernity*, p. 22.
182 Murphy, "Anglo-American Postmodernity: A Response to Clayton and Robbins," *Zygon* 33/3 (Sep. 1998), p. 479. Murphy gives this example: "Saint Francis's goodness supervenes on a collection of his character traits: generosity, chastity, and so forth. Goodness, however, is not *reducible* to this set of character traits because goodness depends, in addition, upon circumstances. For example, Saint Francis's celibacy and his giving away his property might be judged quite differently if he had been a married man with children to support" (ibid.).
183 Murphy, *Beyond Liberalism and Fundamentalism*, p. 138.
184 Murphy, *Beyond Liberalism and Fundamentalism*, p. 138, note 10.
185 Murphy, *Anglo-American Postmodernity*, p. 119.
186 Murphy, *Beyond Liberalism and Fundamentalism*, p. 147.

The Presence of the Triune God

211

ahead involves an emphasis on God's immanence at the quantum level."[187]

Murphy identifies two prominent strategies among those who seek to integrate scientific and theological accounts of divine action. The first of these approaches is "Peacocke's claim that God works solely in a top-down manner, influencing the whole of the universe in a way analogous to the way the environment influences an organism, or as the whole person influences his or her own bodily actions."[188] She explains a common misunderstanding related to the top-down causation as follows: "to place theology as the top-most science in the hierarchy is not necessarily to say that theology is merely the science of the whole of the empirical order. Rather, it is to say that the behavior of the created universe cannot be explained apart from its relation to an additional kind of reality, namely, God."[189] Then, she indicates some limitations with the application of top-down causation alone: "it is the analogy relating levels of causal description (theology is to science in general as higher-level science is to lower) rather than an analogy between God and natural causes (e.g., the human person) that is useful here."[190] Thus, Murphy presents serious reservations about the adequacy of an account of divine action in terms of top-down causation alone. She admits Peacocke's rejection of any dualistic account. Yet, she argues that, if we understand mental events as a function of the operation of the organism at a high level of organization as in Peacocke, "God would then be the world-mind or the world-soul."[191] She questions this analogy. She claims that, "Since God has no body, we get no help with the question of *how* God brings it about that events obey his

---

187 Murphy, *Beyond Liberalism and Fundamentalism*, p. 148, note 22.

188 Murphy, *Beyond Liberalism and Fundamentalism*, p. 148. Concerning the analogy of the whole person and his or her own bodily actions, she suggests a difficulty "either that God is like the mind or soul of the universe (a dualism that Peacocke rejects), or else a pantheistic view of God and the world" (ibid.). In this case, however, I think the term "panentheistic" is more accurate than "pantheistic" in this case.

189 Murphy, *Beyond Liberalism and Fundamentalism*, p. 149.

190 Murphy, *Beyond Liberalism and Fundamentalism*, p. 149.

191 Murphy, "Divine Action in the Natural Order: Buridan's Ass and Schrödinger's Cat," *Chaos and Complexity: Scientific Perspectives on Divine Action*, eds Robert J. Russell, Nancey Murphy, and Arthur Peacocke (Vatican City State & Berkeley: Vatican Observatory & Center for Theology and the Natural Sciences, 1995), p. 339.

212   *Transcendence and Spatiality of the Triune Creator*

will."[192] Thus Murphy suggests a bottom-up account as a "plausible supplement to Peacocke's top-down approach," stating that, "top-down causation by God should... be expected to be mediated by specific changes in the affected entities, and this returns us to the original question of how and at what level of organization God provides causal input into the system."[193]

The second strategy for giving an account of the "causal joint," Murphy proposes, is to explore quantum physics and seek "to give an account of God's action throughout the natural and human world by means of action at the quantum level (either alone or in conjunction with top-down action)."[194] In her article, "Divine Action in the Natural Order: Buridan's Ass and Schrödinger's Cat," Murphy contends that, in addition to creation and sustenance, there are two modes of God's action within the created order: one at the quantum level and the other through human intelligence and action. She contends, "The apparent random events at the quantum level *all* involve (but are not exhausted by) specific, intentional acts of God."[195] She furthermore raises a provocative question with regard to the quantum level: "Does God produce solely

192 Murphy, "Divine Action in the Natural Order," p. 339.
193 Murphy, "Divine Action in the Natural Order," p. 339.
194 Murphy, *Beyond Liberalism and Fundamentalism*, p. 149. Acknowledging that "[t]hese discussions are promising but as yet inconclusive," Murphy indicates three articles in *Chaos and Complexity* as examples of such discussion: Murphy, "Divine Action in the Natural Order," pp. 325–357, George F. R. Ellis, "Ordinary and Extraordinary Divine Action: The Nexus of Interaction," pp. 359–395, and Thomas F. Tracy, "Particular Providence and the God of Gaps," pp. 289–324. Elsewhere she presents W. G. Pollard as the theorist in this area whose work comes closest to hers. In his book, *Chance and Providence: God's Action in a World Governed by Scientific Law* (New York: Scribner, 1958), Pollard suggests that "God works through manipulation of all sub-atomic events" (N. Murphy, "Divine Action in the Natural Order," p. 355). John Polkinghorne also points to W. G. Pollard as the pioneer who interprets divine action at a quantum level. Even though he does not consent to this strategy, Polkinghorne admits the general popularity of it. Concerning the indeterminacy of the quantum level, Polkinghorne states as follows: "Heisenberg's uncertainty principle, which made the epistemological assertion of the simultaneous unknowability of both position and momentum, has been widely interpreted as a principle of indeterminacy, with the ontological implication that quantum entities do not possess at all times definite position and momenta" [See J. Polkinghorne, *Belief in God in an Age of Science* (New Haven & London: Yale University Press, 1998), p. 53].
195 Murphy, "Divine Action in the Natural Order," p. 339.

The Presence of the Triune God

and directly all the events (phenomena) at this level, or are the entities endowed with 'powers' of their own?" To this question, she presents two possible answers: "either God is the sole actor at this level or the entities (also) have their own (God-given) powers to act." Murphy quickly dismisses the first option on theological grounds:

> To say that each sub-atomic event is solely an act of God would be a version of occasionalism, with all the attendant theological difficulties mentioned above: it exacerbates the problem of evil; it also comes close to pantheism, and conflicts with what I take to be an important aspect of the doctrine of creation – that what God creates has a measure of independent existence relative to God, notwithstanding the fact that God keeps all things in existence.[196]

She emphasizes that "All created entities, despite being sustained by God, are entities in their own right *vis-à-vis* God."[197] Thus, Murphy states, "God governs each event at the quantum level in a way that respects the 'natural rights' of the entities involved."[198] As does Peacocke, Murphy opposes both interventionist accounts and "God of the gaps" accounts of divine action. Interventionism supposes that God should violate the laws of nature he has established. The view of "God of the gaps" has epistemological problems: "science will progress and close the gaps."[199] But Murphy also states a more basic intuition behind the rejection of both of them: "God must not be made a competitor with processes that on other occasions are sufficient in and of themselves to bring about a given effect. In addition, if God's presence is identified with God's efficacy then a God who acts only occasionally is a God who is usually absent."[200]

Murphy asserts that "God's action at the sub-atomic level governs the behavior of nature's most basic constituents, but without violating their

---

196 Murphy, "Divine Action in the Natural Order," p. 340.
197 Murphy, "Divine Action in the Natural Order," p. 342.
198 Murphy, "Divine Action in the Natural Order," p. 343.
199 Murphy, "Divine Action in the Natural Order," p. 343. In the same vein, Polkinghorne defines and dismisses the notion of "God of the gaps" as follows: "The invocation of God as an explanation of last resort to deal with questions of current (often scientific) ignorance. ('Only God can bring life put of inanimate matter,' etc.). Such a notion of deity is theologically inadequate, not least because it is subject to continual decay with the advance of knowledge" (Polkinghorne, *The Faith of a Physicist*, p. 197).
200 Murphy, "Divine Action in the Natural Order," p. 343.

214   *Transcendence and Spatiality of the Triune Creator*

'natural rights.'" At the same time, she claims, "God affects human consciousness by stimulation of neurons – much as a neurologist can affect conscious states by careful electrical stimulation of parts of the brain. God's action on the nervous system would not be from the outside, of course, but by means of bottom-up causation from within."[201] Murphy reminds us of the fact that "medieval mystics placed a great deal of emphasis on the faculty of *memory* – taking it as an important means by which revelation is formed of materials available in the person's culture."[202]

Concerning Murphy's proposal that God interacts within this world by manipulations at the quantum level, Polkinghorne raises the objection that "[s]ubatomic events scarcely look like promising locations for holistic causality. After all, one could hardly get more 'bits and pieces' than elementary particles."[203] He states, "if quantum theory does have a role to play in solving the problem of agency, it will only be because its effects are amplified in some way to produce an openness at the level of classical physics."[204] Thus Polkinghorne asserts that the proponents of the view that divine action takes place through quantum events do not seem to have been able to articulate a clear account of how this could actually be conceived as the effective locus of providential interaction.

Polkinghorne maintains that there are two possible sources of openness in physical process: quantum theory and chaos theory. For him, unlike Murphy and other supporters of quantum option, however, quantum theory has a serious difficulty for space, room, or "gaps" of divine action. Polkinghorne states that, "The unpredictabilities of quantum processes relate solely to the outcomes of measurements (of course, not necessarily consciously observed). The proposal is that God should de-

201 Murphy, "Divine Action in the Natural Order," p. 349.
202 Murphy, "Divine Action in the Natural Order," p. 350. As an example, Murphy presents the following illustration: "A student reported that the thought suddenly occurred to him that he should speak to a recent acquaintance about his drinking problem – even though he did not know that the person had such a problem. In conjunction with the thought, he had a sudden memory of his troubled relationship with his father due to alcohol, and felt an associated emotional impact from that memory. The conjunction of all of these experiences convinced him that he was receiving a message from God to approach the acquaintance and urge him to the alcohol problem before it affected his relations with his children" (ibid.).
203 Polkinghorne, *Belief in God in an Age of Science*, p. 60.
204 Polkinghorne, *Belief in God in an Age of Science*, p. 60.

The Presence of the Triune God

termine some, at least, of these results. In between measurements, quantum theory is perfectly deterministic."[205] In this view, he comments, divine action is not continuous but episodic. For Polkinghorne, quantum theory offers no more than a small part of the story of divine action.[206]

As opposed to quantum theory, Polkinghorne proposes chaos theory as a way of thinking about human and divine agency. Concerning chaos theory, he explains as follows:

> Physical systems which are non-linear (doubling the cause produces something more complicated than just twice the effect) and reflexive (the system can act back upon itself), often display such an exquisite sensitivity to their detailed circumstance that their behaviour becomes intrinsically unpredictable. The way they behave is not unrestrictedly haphazard, however, but they explore a confined range of possibilities (called a 'strange attractor'). Thus chaos systems exhibit a kind of ordered disorder. The recognition of this behaviour is comparatively recent, and the study of chaos theory is still at an early stage of development. Some regard it as a 'third revolution' in physics, comparable to the discoveries of Newtonian mechanics and quantum theory.[207]

205 Polkinghorne, *Science and Theology: An Introduction* (London/Minneapolis: SPCK/Fortress, 1998), p. 89. Elsewhere he comments on quantum phenomena in the same vein: "It is notorious that quantum events are believed by the majority of physicists to be constrained only by overall statistical regularity in their patterns of occurrence. Individual events are characterized by a radical randomness and are even spoken of as being 'uncaused'. It might be thought that here is to be found the necessary room for manoeuvre, both for God and for ourselves. Such a view has been proposed but it has not commended itself widely. It is likely to founder on the propensity for randomness to generate regularity, for order to arise from chaos. The aggregation of individually chance events at one level is liable to compose itself into a highly predictable pattern at a higher level. The practice of Life Insurance Offices is based upon this very tendency" [*Science and Providence: God's Interaction with the World* (London: SPCK, 1989), p. 27].

206 Polkinghorne, *The Faith of a Physicist: Reflections of a Bottom-Up Thinker* (Minneapolis: Fortress, 1996), p. 25. Concerning Peacocke's view on quantum alternatives, Ian Barbour comments as follows: "Arthur Peacocke takes quantum effects to be only one example of chance, which occurs at many points in nature. Moreover, he portrays God as acting through the whole process of chance and law, not primarily through chance events. God does not predetermine and control all events; chance is real for God as it is for us. The creative process is itself God's action in the world." See I. Barbour, *Religion and Science: Historical Contemporary Issues* (New York: HarperCollins, 1997), p. 188.

207 Polkinghorne, *The Faith of a Physicist*, p. 196. Ian Barbour explains chaos theory as follows: "The theory of nonlinear dynamic systems in which infinitesimal

216   *Transcendence and Spatiality of the Triune Creator*

Polkinghorne claims that the intrinsic unpredictabilities of the physical world are found not only in the quantum world, but, since the emergence of chaos theory, have been recognized also in the everyday classical world as well. He states that, "The discovery of chaotic systems, whose exquisite sensitivity to the fineness of detail of their circumstances imposes severe restrictions on predictability, even in the regime of classical Newtonian physics."[208]

Polkinghorne asserts that there is an inescapable need to discuss chaotic systems in a holistic context. He writes, "the chaotic systems can never be isolated from their environment because of their extreme sensitivity to any external disturbance, however minimal."[209] To make clearer how chaos theory works, Polkinghorne adopts the concept of "active information": "'active,' because the holistic principle brings about actual future behaviour; 'information,' because its action relates to structure rather than to energetic properties."[210] This concept of top-down-causality by active information, according to him, supplements the bottom-up causality of energetic interchange between constituents. Thus, he argues, "The whole is indeed more than the sum of its parts because it exerts an influence on the parts. There is causality that flows from top to bottom, as well as from bottom to top."[211]

Murphy acknowledges only a subsidiary role of chaos theory for explaining divine action in the physical world. She states that, "The real

---

changes in initial conditions can result in very large changes in subsequent behavior" [*Religion and Science: Historical and Contemporary Issue* (New York: Harperollins, 1997), p. 357]. He refers to the so-called "butterfly effect": "In chaotic systems, an infinitesimally small uncertainty concerning initial conditions can lead to enormous uncertainties in predicting subsequent behavior. This has been called 'the butterfly effect' because a butterfly in Brazil might alter the weather a month later in New York. The effect of moving an electron on a distant galaxy might be amplified over a long period of time to alter events in the earth" (ibid., p. 183).

208 Polkinghorne, *Scientists as Theologians: A Comparison of the Writings of Ian Barbour, Arthur Peacocke and John Polkinghorne* (London: SPCK, 1996), p. 34

209 Polkinghorne, *Scientists as Theologians*, p. 36.

210 Polkinghorne, *Scientists as Theologians*, p. 36. Elsewhere he explains as follows: "Since chaotic systems are unisolatable, the new causal principles would have an holistic character. The term 'active information' has been coined to describe this new kind of causality ('active' because of its causal efficacy; 'information' because it concerns the formation of patterns of behavior)" (*Science and Theology*, p. 42f).

211 Polkinghorne, *Scientists as Theologians*, p. 36.

The Presence of the Triune God

value of chaos theory for an account of divine action is that it gives God a great deal of 'room' in which to effect specific outcomes without destroying our ability to believe in the natural causal order. The room God needs is not space to work within a causally determined order – ontological room – but rather room to work within our perceptions of natural order – epistemological room."[212] Thus Murphy casts doubt on the propriety of chaos theory: "what chaos shows is not that there is genuine indeterminacy in the universe, but rather that we have to make a more careful distinction between predictability (an epistemological concept) and causal determinism (an ontological concept)."[213] In this context, proposing that "two of Christians' most common subjects for prayer are health and weather," she asks this question, "do we pray for these things rather than others because we lack faith that God could 'break a law of nature' or is it rather because of our long experience with a God who prefers to work on our behalf 'under the cover of chaos'?"[214]

In several places, Polkinghorne presents the question of how God relates to a temporal creation.[215] He states, "A bridge between the scientific and theological discussion can be provided by a topic in which they both have an interest: time."[216] Augustine and Boetius speak for the traditional view of time: "God perceives the whole of cosmic history 'at once,' in a timeless act of knowing by the One who is outside time altogether." Polkinghorne proposes this interpretation of time is, though not necessarily, related to the deterministic character of physical processes of the universe. He comments, "there is a certain degree of alogical association between an atemporal view and determinism, and between temporality and openness."[217] Polkinghorne criticizes the atemporal view, arguing that "the God who simply surveys spacetime from an eternal viewpoint is the God of deism, whose unitary act is that frozen pattern of being."[218] Quoting Barth's saying that "Without God's complete temporality the

---

212 Murphy, "Divine Action in the Natural Order," p. 348.

213 Murphy, "Divine Action in the Natural Order," p. 328.

214 Murphy, "Divine Action in the Natural Order," p. 348f.

215 Polkinghorne, *Science and Theology*, pp. 90–92; *Belief in God in an Age of Science*, pp. 67–69; *Science and Providence*, pp. 77– 84; *Scientists as Theologians*, p. 41.

216 Polkinghorne, *Belief in God in an Age of Science*, p. 67.

217 Polkinghorne, *Science and Theology*, p. 91.

218 Polkinghorne, *Science and Providence*, p. 79.

218 *Transcendence and Spatiality of the Triune Creator*

content of the Christian message has no shape," Polkinghorne claims that "the Christian gospel is an unfolding drama of redemption, not a timeless moment of illumination."[219]

As "one difficulty which all talk of time within God must face," Polkinghorne raises the question "which time?"[220] He contends: "The theory of special relativity abolished the Newtonian idea, seemingly so congenial to our everyday experience, of one, uniformly flowing, universal time. It replaced it by a multitude of individual times experienced by different observers, according to their different states of motion."[221] In other words, relativity theory recognizes the observer-related character of temporal experience. Thus concerning divine temporality, Polkinghorne asserts that, "God is not a localized observer, but everywhere present. Whatever the divine time axis might be, the frame of God's omnipresence will sweep out the whole of cosmic history, so that God experiences every event as and when it happens."[222]

Like Tracy, as mentioned in the previous chapter, Polkinghorne claims that a discussion of divine temporality is needed for any accounting of divine action. It is regrettable, however, that neither Polkinghorne nor Tracy sees the parallel need to speak of divine spatiality. Nevertheless, against the atemporalist view, Polkinghorne states, "In talk of this kind, time is being assimilated to space, so that the complete history of the universe is thought of as laid out on a four-dimensional space-time 'map' for instant perusal by God."[223] Even in the temporalist view, I think, time is to be assimilated to space. Thus, in the discussion on divine action, I propose we should presuppose some kind of divine spatial-

---

219 Polkinghorne, *Science and Providence*, p. 80. He cites Barth from Gunton's book, *Being and Becoming* (Oxford University Press, 1978), p. 180. One of the problems in Polkinghorne's thought, for me, is that he admits the notion of process theologians' divine dipolarity. Polkinghorne contends, "There must be both temporal and eternal poles to divinity. This notion of divine dipolarity is a gift from the process theologians that has found much wider acceptance among many twentieth-century theological thinkers" (*Belief in God in an Age of Science*, p. 69). See also *Scientists as Theologians*, p. 41 and *Science and Providence*, p. 80.

220 Polkinghorne, *Science and Providence*, p. 81. Elsewhere, he states this question like this: "What time is God's time or, more technically, what is God's frame of reference?" (*Science and Theology*, p. 91).

221 Polkinghorne, *Science and Providence*, p. 81.

222 Polkinghorne, *Science and Theology*, p. 91.

223 Polkinghorne, *Science and Providence*, p. 81.

The Presence of the Triune God 219

ity. Of course, as with the discussion of time, the crucial question here is "which space?" or "what space is God's space?"

One of the important reasons why both Polkinghorne and Tracy are reluctant to refer to divine spatiality is that, they think, it is related to the problem of divine corporeality, or God's embodiment. To uphold the freedom of God, Tracy denies the embodiment of God. He thinks of embodiment as a limitation on God. Since God is a perfect agency, Tracy asserts, God is an unembodied being.[224] Likewise, by observing four characteristics of human bodily experience, Polkinghorne presents difficulties with the analogy of embodiment:

> 1 Some power of direct action (I can move my limbs, but I cannot directly will a change in the peristaltic rhythms of my intestines). 2 Some degree of direct awareness (I feel aches and pains but I am not directly aware of my blood pressure). 3 A limitation of perspective (I view the world from 'within' my body). 4 A vulnerability to changes in the physical circumstances of my body (if my brain tissues degenerate because of Alzheimer's disease, my personality will become demented).[225]

According to Polkinghorne, the analogues of 1 and 2 might be acceptable, but only if they were suitably stretched and enhanced by the replacement of partial power by total ability. They would then provide for comparison with divine omnipotence and omniscience in relation to the world. However, "limitation 3 would be unacceptable for God."[226] Finally, for Polkinghorne, property 4 is the rock on which the idea of divine embodiment finally founders. On this note, Polkinghorne agrees with Vanstone's description of the Christian God as "the God who is impassible but vulnerable." However, he opposes Jantzen's view that "if the universe is the embodiment of God, then the universe taken as a whole must be everlasting." Polkinghorne states, "Even if the idea of cosmic oscillation were correct, combined with divine embodiment it

---

224 See Thomas F. Tracy, *God, Action and Embodiment* (Grand Rapids: Eerdmans, 1984).

225 Polkinghorne, *Science and Providence*, p. 18f.

226 Polkinghorne cites Richard Swinburne who makes this a major ground for the criticism of embodiment ideas. Then, he acknowledges that "Grace Jantzen, who is the most fluent contemporary defender of divine embodiment, has grounds for replying to this by saying: 'But this latter consideration does not, surely, count against God's having a body but in favour of it – only, his body must be understood to be the whole universe, not an individual part of it'" (*Science and Providence*, p. 19).

220   *Transcendence and Spatiality of the Triune Creator*

would have the curious theological consequence of a God everlastingly engaged in ceaseless change. He would be the Cosmic Quick Change Artist. Such a God could hardly be called impassible."[227] Polkinghorne maintains that the analogy of divine embodiment suggests that God must destroy the liberty of creation if he is to safeguard his own independence. For, according to Polkinghorne, "God and the world are so closely linked by embodiment that one must gain the mastery over the other." Thus, he holds, "Only breaking the implicit in embodiment can God be let be to be God and his creation be let be to be itself."[228] However, abandonment of the notion of God's embodiment, or corporeality, in this universe, Polkinghorne asserts, does not mean that we must give up any hope of gaining analogical insight into his action in the world from the consideration of our own embodiment.[229] Nevertheless, emphasizing the difference between God and human beings, he comments on the impropriety of God's embodiment as follows:

> The most important thing is that *we* are constituted by our physical bodies and so are in thrall to them. Their decay is our dissolution, though not without the Christian hope of a destiny beyond that dissolution through God's act of resurrection, reconstituting us in a new environment of his choosing. God, on the other hand, is not constituted by the cosmos, even in part of his nature, and so he is never in thrall to it... Though God interacts with the world it is not proper to speak of his being embodied in it.[230]

This view of Polkinghorne against God's embodiment is appropriate. He aptly indicates some difficulties with the contention that the world is God's body. Due to these reasons, however, the reluctance to acknowledge God's spatiality is not proper. While many theologians argue for a conceptualization of divine temporality, fewer contend for divine spatiality. But, we argue, it is not reasonable to oppose God's spatiality as a conclusion drawn from the assertion of God's spirituality. Although we need not view the world as the body of God, we do need to account for God's glory being embodied through material things, including the human body. In these processes of embodiment, God uses space as his vehicle or medium. Dyrness's third category, "embodiment," need not im-

---

227   Polkinghorne, *Science and Providence*, p. 21.
228   Polkinghorne, *Science and Providence*, p. 22.
229   Polkinghorne, *Science and Providence*, p. 23.
230   Polkinghorne, *Science and Providence*, p. 34.

The Presence of the Triune God 221

ply that the world is the body of God, but rather that God embodies himself in the world in various ways through material things.

4. Embodiment

This section will examine modern theological discussion on God's embodiment as Dyrness's third category. God is not a bodily being. Nor does he have his own body. I also reject the view of the world as the body of God. But I will argue that God embodies himself in various ways. Creation is not just the background of our salvation but is the theater for God's glory. Creation is the place for the embodiment of God's purposes (subsection 1). The creator God has come into the creation that he has made. In the incarnation of the Son, God has embodied himself in space and time. The incarnation is the incarnate embodiment of the fullness of God in Christ. It is not a temporary episode but a trinitarian event (subsection 2). The church of both the Old and New Testaments is in an important sense the earthly-historical form of God's own existence. The church is the congregation of people of God, the body of Christ, and the temple of the Holy Spirit. God's people embody God in this world in a quite literal manner. She can be in the life of the triune God by means of the Holy Spirit (subsection 3). The Eucharist is the visual and material means of God's invisible grace. The whole Christ is present in the Eucharist by the Holy Spirit. The Lord who is present in the Eucharist will come to accomplish the new creation that he has already begun in the days of his flesh (subsection 4). New creation is the work of making the creation a suitable vehicle for God's glory. It will be an embodied event. The new creation is not a total annihilation of creation. It is the transformation of the world of creation. God will be present in this new creation: "See, the home of God is among mortals" (Rev. 21:3) (subsection 5). God uses space, time and bodies as vehicles of his grace. By means of them, God expresses his goodness and love for his creatures. Thus, my position will reject the idea of the literal embodiment of God in the world, while acknowledging a kind of divine embodiment. In processes of embodiment, God uses the material beings, including the human body, as vehicles for his embodiment. For this to be possible,

222   *Transcendence and Spatiality of the Triune Creator*

God must have his own spatiality. This is the third and last meaning of God's spatiality.

4.1 Creation[231]

Dyrness claims that some of the reasons for "the eclipse of creation" lie in the way that theology has been approached since the Reformation, "either as a system of true knowledge inwardly grasped, or, more recently, as a narrated story of God's acts in history."[232] Dyrness cites the case of Gerhard von Rad as an example. According to von Rad, we begin to read the Hebrew scriptures at Deuteronomy 26:5. In that verse, we find the major theme of the first six books of the Old Testaments, that is, Israel's gratitude for God's deliverance from Egypt. Thus, von Rad considers the creation account a minor later addition to that theme, or as a theological after-thought. In von Rad's interpretation, creation is eclipsed as a kind of backdrop to the "important stuff" of salvation history. Dyrness comments on von Rad's view on creation: "It is the confession in Deuteronomy that answers the question of what God is going to do about this marvelous world that is so scarred. Meanwhile, the creation accounts simply 'undergird this faith by the testimony that this Yahweh, who made a covenant with Abraham and at Sinai, is also the creator of the world.'"[233]

---

231   Pannenberg rightly indicates the distinction between "creation as an act of God" and "creatures as the products of divine activity" in the traditional doctrine of creation (Pannenberg, *Toward a Theology of Nature*, p. 34). Regardless of their difference, however, I use the term of "creation" in this subsection, which is in relation to the last subsection "new creation."

232   Dyrness, *The Earth is God's*, p. 27.

233   Dyrness, *The Earth is God's*, p. 27. This quotation comes from Gerhard von Rad., *Genesis*, ed. and trans. John Mark (Philadelphia: Westminster, 1961), pp. 43, 44. Cf. Bernhard W. Anderson, *From Creation to New Creation* (Minneapolis: Fortress Press, 1994), p. vii. In his editor's foreword, Walter Brueggemann indicates the "eclipse" of creation in Old Testament studies. Like Dyrness, he also conceives of the case of Gerhard von Rad as the reason for that eclipse. However, Brueggemann limits his charge against von Rad to his early writings. He states as follows: "The "eclipse" of creation in Old Testament studies has been largely determined by the categories of Gerhard von Rad, who, at least in his early writings, made creation subordinate to salvation history. In his latest work, however, even von Rad was changing his mind about the matter in significant ways."

The Presence of the Triune God 223

As another cause of creation's being reduced to the study of "beginnings," Dyrness points to the evolutionary view of human origins. This secular discussion is limited to the question of process: "How did the world come to be the way it is?" Yet, Dyrness observes that "The biblical accounts of creation are clearly more interested in why God made the world and what the divine purposes are for its continued existence than in the process of creation."[234] In an overreaction to the evolution, we have missed the relevant point of the creation accounts in the Bible. Additionally, Dyrness refers to "the preference for theology as sure knowledge." This bias is another modern reason for the eclipse of interest in creation. The preference for theology as sure knowledge, according to him, has led us "to the privileging of the oral over the visual and the abstract concept over image and metaphor." Following Gunton, Dyrness maintains that "it was Hegel in the last century who crystallized the idea that comes ultimately from Plato, that concepts are alone reliable because they are purged from visual imagery which is inherently unreliable."[235] In this context, Dyrness argues for the metaphorical character of biblical language. "Metaphor, by a transfer of contexts," Dyrness writes, "helps us 'see' things that in their ordinary context are invisible to us."[236] In an endnote, Dyrness quotes LaCugna's saying: "The systematic theologians need to keep in mind that every concept, whether it be 'substance' or 'relation,' is fundamentally metaphorical, not a literal description of what is."[237]

For Dyrness, creation is the place for a kind of divine embodiment, i.e., concrete manifestations of God's eternal purposes. For his purposes, God made use of the covenant idea. The covenant is a metaphor that God uses to help us understand the relationship God wished to establish with the chosen people. The creation account is being placed at the beginning of the Hebrew scriptures, Dyrness emphasizes "God's intention that Is-

---

234 Dyrness, *The Earth is God's*, p. 28. Concerning the question "whether or not the Bible allows for belief in evolution," Dyrness contends that, "Clearly that is asking the text a question that it does not intend to answer" (ibid.).

235 Dyrness, *The Earth is God's*, p. 28f.

236 Dyrness, *The Earth is God's*, p. 29. Even though "metaphors do not tell the whole truth" and "the problem comes when we take metaphors for the whole truth," unlike Sallie McFague or Elizabeth Johnson, Dyrness contends, "the part they reveal turns out to be essential" (ibid.).

237 LaCugna, *God for Us*, p. 359, quoted by Dyrness, *The Earth is God's*, p. 172, note 14.

224   *Transcendence and Spatiality of the Triune Creator*

rael understand itself in relationship both to the physical creation and to the nations of the world." Quoting Brian Walsh and Richard Middleton ("What this canonical placing of creation does is decisively to reinterpret the Exodus-Sinai story in terms of a larger more comprehensive metanarrative"), Dyrness maintains that placing the creation account as the prologue of all God's work "underlines the fact that history must be understood in terms of creation, time in terms of space."[238]

Moltmann presents a criticism that "Barth connected creation and covenant together so exclusively as a pair, and [...] did not revise the anthropocentrism in pairing creation and covenant which he in fact shares with 'modern religion.'"[239] Barth relates God's covenant only to human beings. In this sense, Barth's doctrine of creation takes on an anthropocentric focus. Thus Barth says that the covenant may be regarded as "the internal basis of the creation" and the creation may be regarded as "the external basis of the covenant."[240] For Barth, Moltmann argues, "Jesus Christ is the beginning, middle and end of the covenant of God with humankind, as though Christ had come only for their sake and were only their reconciler and Lord. The creation of the non-human world is explained as the 'external basis of the covenant.'"[241] By using Barth's own words, however, Moltmann revises Barth's view: "'Creation itself is the beginning of this history' (viz. of the covenant). If that is the case, then the creation itself is already God's covenant. If creation as a whole is God's covenant from the beginning, then all creatures live in this covenant, and not just human beings, and as 'centre of the covenant.'"[242] Moltmann claims that not the "historicizing of nature" but

---

238   Dyrness, *The Earth is God's*, p. 33. This citation comes from B. Walsh and R. Middleton, "Facing the Postmodern Scalpel: Can the Christian Faith Withstand Deconstruction?" in *Christian Apologetics in a Postmodern World*, eds Timothy and Dennis L. Okholm (Downers Grove, IL: InterVarsity Press, 1995), p. 148.

239   J. Moltmann, *History and the Triune God: Contributions to Trinitarian Theology* (New York: Crossroad, 1992), p. 129.

240   K. Barth, *Church Dogmatics* III/1, pp. 42ff, and 338ff.

241   Moltmann, *History and the Triune God*, p. 128.

242   Moltmann, *History and the Triune God*, p. 129. Contrary to Moltmann, though he acknowledges its problems, Gunton, "The Doctrine of Creation, p. 154, interprets Barth's view as follows: "This link between creation and redemption, despite its problems, enabled the two to come into far more positive relation than had sometimes been the case. It also enabled Barth to develop fine accounts of providence, of the human person and of the ethics of creation."

The Presence of the Triune God

rather the eschatological "naturalizing of history" is the goal of creation. He writes, "The creation, namely the creation transformed by the indwelling of God, is the goal of history."[243]

As is well known, the traditional Christian doctrine of creation was born in its struggle against the Gnostics. As a definition of the term "Gnosticism," David F. Ford states, "Name (derived from the Greek word *gnosis*, knowledge) given to a varied and diffuse religious movement which saw creation as the work of an inferior god, sharply divided the physical from the spiritual, and offered to initiate exclusive knowledge enabling them to escape from the world and physicality to union with the supreme divine being."[244] Since the early church, the gospel of Christianity has been influenced and threatened by a wide variety of Gnostic-like heresies. A corollary result of this influence is the denegation of the physical, or visible, world of creation as inferior to the spiritual, or invisible, one. Irenaeus' main theological foes were the Gnostics.[245] Against them, he contended for salvation as the recapitulation of God's creative work, not as separation from creation. This view of Irenaeus on creation was inherited by Calvin. Calvin argued that creation is the theater for the display of God's glory. Calvin went so far in one place as to say, "I confess, of course, that it can be said reverently, provided that it proceeds from a reverent mind, that nature is God."[246]

Colin Gunton rightly states that "the only satisfactory account of the relation between the creator and the creation is a trinitarian one."[247] "By strengthening the trinitarian aspects of the doctrine of creation," Gunton states, "Irenaeus was able, first, to develop a markedly positive view of

---

243 Moltmann, *History and the Triune God*, p. 129.

244 Ford, ed., *The Modern Theologians*, p. 741f.

245 Cf. Gunton, "The Doctrine of Creation," Gunton, ed., *The Cambridge Companion to Christian Doctrine* (Cambridge University Press, 1997), p. 148, writes, "he [Ireneus] was helped by his need to out-think the strongly dualistic philosophy of some of his Gnostic opponents."

246 Dyrness, *The Earth is God's*, p. 33. This quotation comes from Calvin, *Institutes*, 1.5.5. See also Bouma-Prediger, *The Greening of Theology*, p. 86. According to Bouma-Prediger, while Calvin immediately qualifies this statement, it is nonetheless quite surprising. Dyrness further clarifies this qualification of Calvin. He states, "Calvin goes on to qualify the statement, showing that it is the sovereignty of God that he wishes to underline" (*The Earth is God's*, p. 171).

247 C. E. Gunton, *Christ and Creation* (Carlisle, 1992), p. 75. T. F. Torrance, *The Christian Doctrine of God*, p. 212, agrees with Gunton in this point.

226   *Transcendence and Spatiality of the Triune Creator*

the value of the created order, material and spiritual alike."[248] Gunton likes Irenaeus' view of creation as the "two hands" of God, the Son and the Spirit. Thus, Gunton holds, "God's transcendence as the maker of all things is not such a kind that he is unable also to be immanent in it through his 'two hands.'"[249]

Creation is not just the background of our salvation. In this sense, Barth's saying that the creation is the external basis of the covenant is not the whole story. Creation itself is a covenant, as in Moltmann's description, "the covenant of creation."[250] After the completion of the creation, God saw his creation and said, "It is very good." Unlike the Gnostic view, the Christian doctrine of creation tells us that creation is not a necessary evil but the place for the embodiment of God's purpose, eventually for the embodiment of God in the incarnation. Creation constitutes the *theatrum gloriae Dei*, as Calvin has fittingly expressed. The goal of history is not an annihilation of creation, but creation transformed by the indwelling of God, i.e., the new creation.

Concerning the relationship between God and space-time in relation to the doctrine of creation, Torrance argues that, though spatial and temporal relations are produced through God's creation of the universe and maintained through His interaction with what God has made, the doctrine of creation does not mean that God stands in a spatial or temporal relation to the creation. Since "all creation is comprehended by God and endowed with rationality in this way," Torrance holds, "space and time are to be conceived not only as relations arising in and with created existence but as bearers of its immanent order."[251] This relation between God and space-time is not contradicted by the incarnation. For Torrance, the doctrine of the incarnation means that God himself has entered into our world in Jesus Christ through the assumption of a physical body in space and time. Thus, he argues, while it does not discount the absolute priority of God over all space and time, the doctrine of the incarnation "asserts the reality of space and time for God in His relations with us and binds

---

248   Gunton, "The Doctrine of Creation," p. 148.
249   Gunton, "The Doctrine of Creation," p. 142.
250   Moltmann, *History and the Triune God*, p. 128, asserts that "creation is not just a work of God which makes the covenant possible but is itself a covenant with God, the covenant of creation." In this context, Moltmann introduces the view of Cocceius, a Reformed federal theologian.
251   Torrance, *Space, Time and Incarnation* (Edinburgh: T&T Clark, 1969), p. 23f.

The Presence of the Triune God

227

us to space and time in all our relations with Him."[252] The creator God has come into the creation which he has made and we encounter God there.

4.2 Incarnation

God wants creation to be the theater of his glory. In the incarnation God becomes a part of the creation. The incarnation of the Son of God, in Dyrness's words, is "the central way that God embodied his glory in creation."[253] The incarnation thus serves not only for the redemption of the fallen world but also for the accomplishment of the divine presence. Thus, Dyrness states, "From the beginning of Jesus' ministry the sense that God was present in creation in a new and more intimate way was palpable."[254] Dyrness's view sounds a lot like Torrance's saying that, by the incarnation, although he was not far off before, the Son of God has come into our spatial realm.[255] The Son of God is always present in this world of creation. By the incarnation, however, the Word of God made flesh became present in it in a new and more intimate way. According to Torrance, this understanding of the Christian doctrines of creation and incarnation was undoubtedly offensive to the Greek mind and conflicted sharply with Greek philosophical categories of the necessity, immobility and impassibility of God.[256]

From the perspective of Pannenberg's "Christology from below," if we speak of the descent into human flesh of a pre-existent Logos, the doctrine of the incarnation is a mythical conception which needs demythologizing in a Bultmannian sense. Gunton reasserts this point: In other

---

252 Torrance, *Space, Time and Incarnation*, p. 24. Torrance further argues, "Thus the miraculous activity of God in the Incarnation is not to be thought of as an intrusion into the creation or as an abrogation of its space-time structure, but as the chosen form of God's interaction with nature in which He establishes an intimate relation between creaturely human being and Himself. Here space and time provide the relational medium within which God makes Himself present and known to us, and our knowledge of Him may be grounded objectively in God's own transcendent rationality" (ibid.).
253 Dyrness, *The Earth is God's*, p. 51.
254 Dyrness, *The Earth is God's*, p. 51f.
255 Torrance, *Space, Time and Resurrection*, p. 14.
256 Torrance, *Trinitarian Faith*, p. 89.

228 *Transcendence and Spatiality of the Triune Creator*

words, Pannenberg associates "a christological *method* from above with the *structure* of concepts which speak of 'the descent of the Son from the world above' [...] Pannenberg's source is surely Bultmann's celebrated attribution to the New Testament writers of a belief in a 'three-decker universe' so different from our own cosmology."[257] Gunton presents two of many assumptions of Bultmann's programme of demythologization. The first assumption is that there is a radical discontinuity between our culture and that of the writers of the New Testament and creeds. Basic to this assumption is that the early Christology took its form from a Gnostic myth of a divine redeemer coming down to earth from above. This assumption, however, is now widely questioned, according to Gunton, "largely on the ground that the Gnostic myths developed after, not before, early Christology."[258]

Bultmann's second assumption is that the modern world is what it is by virtue of its view of the world as a closed system of cause and effect. Gunton translates the second assumption as follows: "to be modern means to share the Kantian view of space and time as part of the unchanging mental equipment of every human being."[259] The second assumption of Bultmann's programme is grounded upon the absolute concept of space, which Newton and Kant contended for. According to Gunton, this view runs up against "the true problem of space in Christology." Thus, he aptly points out, "*It* [the problem] *is not a question of descent but of co-presence in space.* In the same piece of space there are held to be present both God and man, without loss of the deity of the one or the humanity of the other."[260] Gunton contends, "The debate centred on Bultmann's account has [...] served to obscure the true problem of space in Christology."[261]

Gunton asserts that "many difficulties with transcendence derive from the way in which visual patterns of perception dominate our understanding of space."[262] The visual patterns of perception have the charac-

---

257 C. E. Gunton, *Yesterday and Today: A Study of Continuities in Christology* (Grand Rapids: Eerdmans, 1983), p. 111f. He cites Pannenberg, *Jesus – God and Man* (SCM Press, 1968), p. 33.
258 Gunton, *Yesterday and Today*, p. 112.
259 Gunton, *Yesterday and Today*, p. 112.
260 Gunton, *Yesterday and Today*, p. 114.
261 Gunton, *Yesterday and Today*, p. 113f.
262 Gunton, *Yesterday and Today*, p. 114.

The Presence of the Triune God

teristic of a mutual exclusiveness. They presuppose the absolute notion of space. Gunton writes,

> The things we *see* are mutually exclusive. For example, a patch of blue in a painting excludes the presence of any other colour. This mutual exclusiveness encourages us to conceive space as absolute: a thing is there, and cannot *also* be elsewhere. The notion of the universe as a totally closed order follows from this. Accordingly, the fundamental metaphor dominating the way we see reality is that of the machine, with the world as a series of atomic pieces interrelated with each other *only* in a mutually exclusive way. The notion dies hard in our culture, and is another aspect of the deistic and dualistic way in which we view things.[263]

Concerning the christological implications of absolute space, Gunton states, "the tradition's doctrine of the incarnation seems to take the form of an 'intervention' by God who is *outside* of the machine, a crude piece of messing around in the works with a celestial spanner."[264]

Instead of visual patterning of experience which stresses the mutual exclusiveness of things, Gunton proposes the phenomena of hearing. By the example of music, Gunton asserts that we can hear two or more different things taking place together in space. He draws a profound parallel between music and Christology.[265] But Gunton acknowledges the limit of the parallel: "it is only a parallel, for music remains a this-worldly reality for all its anticipations of heaven, while the claim in classical Christology is that in Christ we have the spatial co-presence of the finite and the infinite."[266] Nevertheless, it is still a useful analogy. Gunton introduces Victor Zuckerkandl, who presents an argument which is valuable in showing at least the conceivability of a different conception of space from that typically derived from our experience of things in exclusive

---

263 Gunton, *Yesterday and Today*, p. 114f.
264 Gunton, *Yesterday and Today*, p. 115.
265 Gunton, *Yesterday and Today*, p. 115. Here is Gunton's explanation: "When, for example, the notes of a major triad are played simultaneouly, we hear in one and the same place three tones which retain their identity and create a new reality, the chord. That is in itself an interesting parallel to the statement that in Jesus Christ there are two co-present realities – what Chalcedon called the human and divine natures – which in their association, the 'hypostatic union,' form a new reality which yet does not do away with the specific characteristics of the old: 'without confusion, without change.'" Arthur Peacocke also uses the analogy of music to discuss the relation between God and the world. See A. Peacocke, *Theology for a Scientific Age*, p. 173ff.
266 Gunton, *Yesterday and Today*, p. 115.

230   *Transcendence and Spatiality of the Triune Creator*

juxtaposition. Quoting Zuckerkandl's saying that "The interpenetration of tones in auditory space corresponds to the juxtaposition of colours in visual space," Gunton maintains that "we must extend our conception of what reality is."[267] Gunton indicates that, in connection with his account of space, Zuckerkandl refers to Michael Faraday, one of the shapers of the modern scientific world view. Faraday argues for "a view of atoms as fields of force rather than discrete substances and a theory of the 'mutual penetrability of the atoms.'"[268] Faraday's view is actually closer to the patristic conceptions of cosmology.[269]

For Torrance, "The relation between the actuality of the incarnate Son in space and time and the God from whom He came cannot be spatialized."[270] This relation of God to space became a very pressing problem for the early church. In solving this problem, according to Torrance, the Nicene fathers developed a "relational conception of space."[271] By the help of this concept of space, Torrance claims, it was possible to conceive that the Lord Jesus Christ "was wholly present in the body and yet wholly present everywhere, for He became man without ceasing to be God. He occupied a definite place on earth and in history, yet without leaving His position or seat [...] in relation to the universe as a whole."[272] In other words, Torrance writes, "While He became incarnate within the physical space of the body He assumed, Christ was not confined or circumscribed by it. He thus became man without leaving the bosom of the Father, and while He became flesh He did not abandon His own immateriality."[273] Without ceasing to be the eternal invisible God, Torrance claims, in his incarnation the Word of God became visible man.[274] The

---

267 Gunton, *Yesterday and Today*, p. 116.
268 Gunton, *Yesterday and Today*, p. 117. Gunton cites Victor Zuckerkandl, *Sound and Symbol: Music and the External World*, trans. Willard R. Trask (Routledge & Kegan Paul, 1956); and Michael Faraday, "A Speculation touching Electric Conduction and the Nature of Matter" in *On the Primary Forces of Electricity*, ed. Richard Laming (1838).
269 Cf. Gunton, *Yesterday and Today*, p. 118. Citing Torrance, *Space, Time and Incarnation*, Gunton states that "patristic conceptions of cosmology are nearer to ours than is often assumed."
270 Torrance, *Space, Time and Incarnation*, p. 3.
271 Torrance, *Space, Time and Incarnation*, p. 4.
272 Torrance, *Space, Time and Incarnation*, p. 13.
273 Torrance, *Space, Time and Incarnation*, p. 82.
274 Torrance, *The Christian Doctrine of God*, p. 74.

The Presence of the Triune God   231

incarnation is an even more astounding act than that of the creation of
the universe out of nothing, Torrance asserts, in that "the Creator becomes creature without ceasing to be Creator, the transcendent becomes
contingent without ceasing to be transcendent, the eternal becomes time
without ceasing to be eternal."[275] In this sense, Torrance opposes the Lutheran concept of *kenosis*. Torrance comments, "The self-humiliation of
God in Jesus Christ, his *kenosis* or *tapeinosis*, does not mean the self-
limitation of God or the curtailment of his power, but the staggering
exercise of his power within the limitations of our contingent existence
in space and time."[276]

The incarnation is the incarnate embodiment of the fullness of God in
Christ. By this incarnation, Torrance avers, "God has revealed to us
something of the innermost secret of his own divine personal life not
otherwise possible."[277] As Paul writes in Colossians 2:9, "it is in Christ
that the Godhead in all his fullness dwells embodied," the whole Trinity
has become revealed in the incarnation of the Son Jesus. One step further, "as the Word of God made flesh," Torrance states, "Jesus Christ
embodies in himself not only the exclusive language of God to mankind
but the faithful response in knowledge and obedience of humanity to
God."[278]

In Pannenberg's later writing, we see Pannenberg complement his
Christology "from below," no longer denying Christology "from above."
In *Systematic Theology* he places himself in the tradition of the Alexan-
drian Logos Christology,[279] while still questioning the traditional notion

---

275   Torrance, *The Christian Doctrine of God*, p. 214.
276   Torrance, *The Christian Doctrine of God*, p. 214f.
277   Torrance, *The Christian Doctrine of God*, p. 143.
278   Torrance, *The Christian Doctrine of God*, p. 17.
279   Pannenberg, *Systematic Theology* 2, p. 299. Pannenberg contends that "The linking
      of Adam typology and Logos christology that Irenaeus initiated is found again in
      the 4th century in Athanasius's work *The Incarnation of the Word*, with an empha-
      sis now on the subtler doctrine of the Logos that Alexandrian theology had mean-
      time developed. The function of this linking was no longer anti-Gnostic but apolo-
      getic." Also cf. S. Grenz, *Reason for Hope: the Systematic Theology of Wolfhart
      Pannenberg* (New York: Oxford University Press, 1990), p. 116f, which says,
      "Pannenberg consciously places himself in the tradition of the Alexandrian logos
      Christology. Yet he emphasizes one crucial difference as well. The older tradition
      failed to view the logos title as an expression of the human uniqueness of Jesus in
      terms of his messianic function. This shortcoming, Pannenberg maintains, is over-

232 *Transcendence and Spatiality of the Triune Creator*

of the incarnation. Pannenberg comments on the traditional view of incarnation as follows: "The incarnation could add nothing to God. It seemed totally absurd to Athanasius to ascribe becoming to God. Even in his physical appearing the Son undergoes no change. In the view of Athanasius the biblical statements about God's faithfulness bear witness to his immutability."[280] For Pannenberg, the incarnation is "God's self-actualization in the world."[281] Pannenberg contends that the incarnation is not a secondary phenomenon nor external to the divine essence.[282] Therefore, the incarnation of Jesus converges with the doctrine of the Trinity in that the incarnation is logically related to the self-distinction of the Son from the Father. The self-distinction of the Son from the Father as the basis of creation and incarnation is not external to the divine essence. Thus, Pannenberg asserts, "the incarnation is not just ascribed to the Son, as distinct from the other persons of the Trinity, by external appropriation,"[283] but is also attributed to the Father. In this sense, the incarnation is related to the doctrine of the creation: "The incarnation of the Son

---

come in his emphasis on the connection between the new Adam Christology and Jesus' role as Messiah."

280 Pannenberg, *Systematic Theology* 1, p. 437, note 214. Concerning the traditional doctrine of the incarnation, Timothy Bradshaw avers, "The traditional procedure of 'from above to below' presupposes the Trinity outside the temporal action of God in Jesus and thus renders the incarnation a secondary phenomenon." Thus he asserts that "Pannenberg criticizes the traditional incarnational procedure of firstly positing the eternal God who secondly acts historically in the man Jesus, as anthropomorphic" [T. Bradshaw, *Trinity and Ontology: A Comparative Study of the Theologies of Karl Barth and Wolfhart Pannenberg* (Lewiston, NY: The Edwin Mellen Press, 1988), p. 277].

281 Pannenberg, *Systematic Theology* 2, p. 389. This is the title of section 3 of the chapter 10 of his *Systematic Theology*. Pannenberg also presents this view in the context of the Trinity: "This self-emptying of the Son (Phil. 2:6-7) is also to be understood as the self-actualizing of the deity of the trinitarian God in its relation to the world that comes into being thereby" (Pannenberg, *Systematic Theology* 1, p. 421).

282 Pannenberg, *Systematic Theology* 2, p. 319. Pannenberg writes, "From the standpoint of the eternal Son his identification with the person of Jesus has the form of incarnation. We must not see this, however, as an accidental happening that is external to his eternal essence. It is logically related to his trinitarian self-distinction from the Father. As the free self-distinction of the Son from the Father is the basis of the possibility of all the creaturely reality that is distinct from God, it is also the origin of the incarnation of the Son in Jesus of Nazareth."

283 Pannenberg, *Systematic Theology* 1, p. 328.

The Presence of the Triune God

demonstrates and fulfils the grace of God and his faithfulness to his covenant, and also, in the sense of the Johannine Prologue, to creation."[284] Thus, Grenz states, "as Pannenberg repeatedly asserts, the self-differentiation of Jesus from the Father belongs to the intertrinitarian life as the self-differentiation of the logos (i.e., the Son) from the Father, which in turn is the basis for creation."[285]

Pannenberg convincingly argues that the incarnation is not external to but rather internal and essential to the divine essence and belongs to the whole Trinity. The incarnation is both the self-distinction of the Son from the Father and the self-actualization of God in the world. These contentions provide an amplification of the traditional doctrine of the incarnation. In the Arian controversy, the patristic fathers emphasized *theologia*, rather than *economia*. As LaCugna has pointed out the traditional view implies that the incarnation does not have an enduring significance but only a tentative significance apart from the divine essence. In this sense, Torrance would complement the traditional doctrine by emphasizing that the incarnation is more than "a temporary episode." The incarnation as the assumption of our corporeal existence by the Son of God, for Torrance, is not "only economic." He holds that "the involvement of the Son of God in our human and creaturely being, even after His resurrection, ascension, and parousia, must be maintained without reserve."[286] The Son who ascended to the right hand of the Father is not bodily present in this world. In traditional Christian thought, the church as the community of the disciples of Jesus is called the body of Christ. Christ also promised to be present in the Eucharist. These assertions must clearly play a role in any theological understanding of space. In the following subsections, I will deal with these topics in turn.

4.3 The Church

In the New Testament account, the emergence of the church of Christ in the New Testament is directly related to the outpouring of the Holy Spirit upon Jesus' disciples. This outpouring of the Spirit is called the

---

284 Pannenberg, *Systematic Theology* 1, p. 436.
285 S. Grenz, *Reason for Hope*, p. 115.
286 Torrance, *Space, Time and Incarnation*, p. 4.

234 *Transcendence and Spatiality of the Triune Creator*

"baptism of the Spirit" (Acts 1:5) and "the gift of the Holy Spirit" (Acts 2:38). George E. Ladd states that we can understand the several extensions of Pentecost related in Acts in light of the following verse found in Paul: "For by one Spirit we were all baptized into one body – Jews or Greeks, slave or free – and were all made to drink of one Spirit" (1 Cor. 12:13). Ladd writes, "The baptism with the Spirit is the act of the Holy Spirit joining together into a spiritual unity people of diverse racial extractions and diverse social backgrounds so that they form the body of Christ – the *ekklēsia*."[287] The church is not to be viewed simply as a human fellowship, but as a "creation through the Holy Spirit." The church is an eschatological community destined to experience the eschatological consummation. The church as the Body of Christ is composed of those who are baptized not only with the Holy Spirit but also in the name of the Lord Jesus and the Father. Baptism in the name of Jesus Christ is "the outward sign of admission to the Christian fellowship."[288]

Concerning the church, we find several expressions in the New Testament. First, in relation to the Old Testament, the church is called "people of God." In his last speech of Acts 7, Stephen refers to the Israelites of the Old Testament as "the assembly in the desert" (NIV), or "the congregation in the wilderness" (NASV and NRSV) (Acts 7:38). However, according to Ladd, by concluding that "God is not limited to the temple (7:47)" and that "the possession of the temple did not assure the Jews of correct religion (7:51-53)," Stephen starts Christianity's breach with Judaism.[289] Concerning the Book of Acts, Ladd claims, "The narrative of the first fifteen chapters of Acts shows how a Gentile church arose, free from the Law. The last thirteen chapters tell how a final breach developed between the church and the synagogue."[290] Finally, Paul speaks of the "Israel of God" (Gal. 6:16). Paul identifies the church as "the true spiritual Israel."[291]

Paul refers to the church as both the Body of Christ and the Temple of the Holy Spirit. According to Ladd, the Body of Christ is the most

---

287 George E. Ladd, *A Theology of the New Testament* (Grand Rapids: Eerdmans, 1994, Revised Edition), p. 384. Ladd calls the disciples of the Lord before Pentecost "the embryo church" (ibid.).
288 Ladd, *A Theology of the New Testament*, p. 386.
289 Ladd, *A Theology of the New Testament*, p. 391.
290 Ladd, *A Theology of the New Testament*, p. 393.
291 Ladd, *A Theology of the New Testament*, p. 584.

The Presence of the Triune God

235

distinctive Pauline metaphor for the church. He comments as follows: "Paul never speaks of the church as a body per se; it is the body *in Christ* (Rom. 12.5) or the body *of Christ* (1 Cor. 12:27). As his body, the church is in some sense identified with Christ (1 Cor. 12:12)."[292] Ladd asserts that the primary emphasis of this metaphor is the unity of believers with Christ. Concerning this metaphor, Torrance maintains that Christ has united the church to himself through the Holy Spirit as his Body. Citing Karl Barth, Torrance states, Jesus Christ has sent out the church into history as "the earthly-historical form of his own existence" in the world.[293] In other words, Torrance takes the term of the Body of Christ not merely a figurative expression, but as expressing an ontological reality within humanity, which affects the whole of the human race.[294]

The temple of God is the metaphor that Paul uses to depict the church as the true Israel. The church is the temple of God in which the Holy Spirit dwells (1 Cor. 3:16). According to Ladd, this metaphor has a threefold emphasis: individual (1 Cor. 6:19), corporate (1 Cor. 3:17), and universal (Eph. 2:19-22).[295] By the use of the metaphor of the temple to designate all three dimensions of the church, Paul reinforces the unity of the church in its diversity.[296] Connecting the temple metaphor to the events surrounding the cross and resurrection of Christ and the outpour-

---

292 Ladd, *A Theology of the New Testament*, p. 590.

293 Torrance, *The Christian Doctrine of God*, p. 229.

294 Torrance, *Trinitarian Faith* (Edinburgh: T&T Clark, 1995), p. 9f. Bonhoeffer more strongly emphasizes the visibility of the church as the body of Christ. He contends, "To flee into invisibility is to deny the call. Any community of Jesus which wants to be invisible is no longer a community that follows him" [Dietrich Bonhoeffer, *Discipleship*, trans. B. Green and R. Krauss, *Dietrich Bonhoeffer Works*, vol. 4, eds M. Kuske and Ilse Tödt (Minneapolis: Fortress Press, 2001), p. 113]. "Since Pentecost Jesus Christ lives here on earth in the form of his body, the church-community" (ibid., p. 218). "The body of Christ takes up physical space here on earth" (ibid., p. 225). The body takes up space. Thus Bonhoeffer states, "incarnation does entail the claim to space granted on earth, and anything that takes up space is visible. Thus the body of Jesus Christ can only be visible body, or else it is not a body at all" (ibid., p. 225).

295 Ladd, *A Theology of the New Testament*, p. 585f. Cf. M. Volf, *After Our Likeness: The Church as the Image of the Trinity* (Grand Rapids: Eerdmans, 1998), p. 142, refers to the recent view of Heon-Wook Park that the body of Christ is not conceived organically as the body of the one person, but rather communally as the body of several persons.

296 Ladd, *A Theology of the New Testament*, p. 586.

236 *Transcendence and Spatiality of the Triune Creator*

ing of the Holy Spirit at Pentecost, Dyrness states, "Christ's body was destroyed only to be reconstituted in the resurrection as the Temple in which believers would dwell – a reality that was constituted by the Holy Spirit on the Day of Pentecost."[297] Quoting Moltmann, Dyrness contends that these events are the New Testament doctrine of creation. "[A]s in the Hebrew scriptures," Dyrness claims, "the Temple becomes the microcosm of the larger creation."[298]

Regarding the relation between the incarnation and the church as the body of Christ, Dyrness states: "It is clear from the New Testament that the reality of Christ's incarnation extends into the life of the people of God who live in Christ by the power of the Holy Spirit. God's people are meant to embody God in a quite literal manner."[299] The church is God's people as the embodiment of God's glory. For Dyrness, the church reflects the trinitarian work of God in creation: The incarnation of Christ is that redoing of creation. In the ascension, Christ sits at the place of authority at God's right hand and becomes the source of salvation. Christ performs this role "as the one who is still on the side of creation." Dyrness writes, "Christ thus realizes the glorified state that perfects God's creation [...] The Holy Spirit is poured out (not simply given!) into our hearts as a bearer of this new created nature."[300] Dyrness claims that "the Holy Spirit moves in God's people to join Christ in offering his perfected humanity to God to the praise of God's glory."[301] Paul calls this corporate body of Christ "a new creation" (2 Cor. 5:17), "which is being renewed in righteousness so that it may one day become a perfect embodiment of God's original purposes in creation."[302]

Miroslav Volf presents a trinitarian ecclesiology. He adopts Moltmann's social model of trinitarian relations and applies it to his ecclesiology. He conceives of the church as "the image of the Trinity." Volf states that "The correspondence between the trinitarian and ecclesial relationships is not simply formal. Rather, it is 'ontological' because it is soteriologically grounded [...] The relations between the many in

---

297 Dyrness, *The Earth is God's*, p. 54.
298 Dyrness, *The Earth is God's*, p. 54.
299 Dyrness, *The Earth is God's*, p. 53.
300 Dyrness, *The Earth is God's*, p. 54.
301 Dyrness, *The Earth is God's*, p. 54.
302 Dyrness, *The Earth is God's*, p. 54.

The Presence of the Triune God

the church must reflect the mutual *love* of the divine per-sons."[303] However, he acknowledges that this correspondence has its limits: "Our notions of the triune God are not the triune God, even if God is accessible to us only in these notions."[304] In this sense, Volf aptly indicates, "the ecclesiologically relevant question is how the church is to correspond to the Trinity *within history*."[305] He pursues this correspondence by means of the relationality and the mutual interpenetration of the trinitarian persons. After examining the relational personhood of the trinitarian persons, Volf argues, "Like the divine persons, so also ecclesial persons cannot live in isolation from one another; Christians are constituted as independently believing persons through their relations to other Christians, and they manifest and affirm their own ecclesial personhood in mutual giving and receiving... A Christian lives from and toward others."[306] Volf develops and advances this idea in the discussion on the perichoretic personhood.

"[T]he trinitarian persons are not only interdependent, but also *mutually internal*,"[307] Volf claims. For Volf, perichoresis is thus "the reciprocal interiority of the trinitarian persons."[308] He states, "The unity of the triune God is grounded neither in the numerically identical substance nor in the accidental intentions of the persons, but rather in their mutually interior being."[309] Volf indicates a difference between trinitarian and ecclesial persons as follows: "In a strict sense, there can be no correspondence to the interiority of the divine persons at the human level. Another human self cannot be internal to my own self as subject of ac-

303 M. Volf, *After Our Likeness*, p. 195. He cites Origen's saying that "the church is full of the holy Trinity" (ibid.).
304 Volf, *After Our Likeness*, p. 198.
305 Volf, *After Our Likeness*, p. 200.
306 Volf, *After Our Likeness*, p. 206. But he indicates that there remains the difference between the relations of the ecclesial and trinitarian persons. He states, "It [the difference] consists, first, in the fact that human beings, though determined by one another, *are* not simply communion, as is the Trinity, but rather must always be held together by an implicit or explicit covenant [...] Second, ecclesial communion on this side of God's new creation can correspond to the perfect mutual love of the trinitarian persons only in a broken fashion" (ibid., p. 207).
307 Volf, *After Our Likeness*, p. 208.
308 Volf, *After Our Likeness*, p. 209.
309 Volf, *After Our Likeness*, p. 210.

238 *Transcendence and Spatiality of the Triune Creator*

tion. Human persons are always external to one another *as subjects*."[310] Thus, he holds that "The indwelling of other persons is an exclusive prerogative of God."[311] For Volf, "the divine persons indwell human beings in a qualitatively different way than they do one another." Concerning the mutual indwelling of the Spirit and human beings, he comments as follows:

> If human beings were *personally* interior to the Spirit in the same way the Spirit is personally interior to human beings, the conclusion "the wind blows where it chooses... *so* it is with everyone who is born of the Spirit (John 3:8) would be reversible. But it is not. This personal interiority is one-sided. The Spirit indwells human persons, whereas human beings by contrast *indwell the life-giving ambience of the Spirit*, not the person of the Spirit.[312]

Thus, Volf rightly argues that the mutual internalization of personal characteristics occurring in the church is done through the Holy Spirit indwelling Christians. He asserts, "The Spirit opens them [Christians] to one another and allows them to become *catholic persons* in their uniqueness."[313] Therefore, in spite of the asymmetry of the perichoresis of trinitarian persons and human persons, Volf contends for the interecclesial relevance of the perichoresis of the divine persons. This correspondence between Trinity and church is grounded on the *catholicity* of the divine persons. Volf maintains that, "By opening up to one another both diachronically and synchronically, local churches should enrich one another, thereby increasingly becoming catholic churches."[314] In this way, he asserts, local churches "will also increasingly correspond to the catholicity of the triune God, who has already constituted them as catholic churches, because they are anticipations of the eschatological gathering of the entire people of God."[315]

Quoting the saying of Adrienne von Speyr that "The relationship of the divine persons to one another is so wide that there is room for the

310 Volf, *After Our Likeness*, p. 210f.
311 Volf, *After Our Likeness*, p. 211.
312 Volf, *After Our Likeness*, p. 211.
313 Volf, *After Our Likeness*, p. 212. He states, "Every person is a catholic person insofar as that person reflects in himself or herself in a unique way the entire, complex reality in which the person lives" (ibid.).
314 Volf, *After Our Likeness*, p. 213.
315 Volf, *After Our Likeness*, p. 213.

The Presence of the Triune God

whole world in it," Moltmann holds that "the perichoretic unity of the triune God is an inviting and uniting unity, and as such a unity which is open to human beings and the world."[316] Thus, he does not understand the trinitarian concept of the unity of the triune God to be exclusive, but rather he understands it to be inclusive. By means of his expression "the open Trinity," Moltmann contrasts his position with the traditional figures of the circular, or triangular, Trinity. He states that "The Trinity is 'open' by virtue of its overflowing, gracious love. It is 'open' to its beloved creatures who are found and accepted."[317]

The contentions of Moltmann prompt a question: How might Moltmann evaluate Volf's view on the relationship between the Trinity and ecclesiology? For Moltmann, I suggest that the church in Volf's view is too external to the Trinity. Moltmann's open Trinity includes the whole world, including the church. The triune God and the church do not just correspond to each other, but the relation between them is internal to the Trinity. In this sense, Volf does not fully accept Moltmann's thoughts, at least not in their panentheistic tendencies. If Volf did emphasize the role of the Spirit, I think that he could fully develop his ecclesiology without falling into panentheism. The church can not exist within the Trinity literally or spatially. But she can be in the life of the Triune God by means of the Holy Spirit.

Torrance indicates that from early times, Christianity, especially Latin Christianity, had assimilated the idea of the receptacle into its doctrines of the church, Sacraments. He comments, "Supernatural grace was widely thought of as contained in ecclesiastical vessels and capable of being handed on in space and time by means of them."[318] Unlike in the East where the question of space came to be thought out primarily from a consideration of creation and incarnation, in the West the fundamental approach to the question of space came to be thought out from a consid-

---

316 Moltmann, *History and the Triune God*, p. 87. Elsewhere Moltmann writes, "The trinitarian relationship of the Father, the Son and the Holy Spirit is so wide that the whole creation can find space, time and freedom in it" (*The Trinity and the Kingdom*, p. 109). Moltmann refers to the Holy Spirit as a broad place: "So in the new life we experience the Spirit as a 'broad place' – as the free space for our freedom, as the living space for our lives, as the horizon inviting us to discover life. 'The broad place' is the most hidden and most silent presence of God's Spirit in us and round about us" (*The Spirit of Life*, p. 178).

317 Moltmann, *History and the Triune God*, p. 87.

318 Torrance, *Space, Time and Incarnation*, p. 25.

240 *Transcendence and Spatiality of the Triune Creator*

eration of the real presence of Christ in the Mass. Thus, Torrance states, the receptacle concept of space "led the Medievals to think of the presence of God almost entirely in a *spatial* manner, *apart from time*, which helps to explain the inadequate place they gave in their thought to a dynamic relation between God and historical existence and to the temporal relevance of the Last Things."[319] If this is so, our under-standing of God and time must engage fully with the question of this "real presence." In the next subsection, I will examine the issue of the Eucharist and Christ's presence in it.

4.4 Eucharist

In his book, *The Christian Doctrine of God*, T. F. Torrance points out a distinctive approach that is appropriate to the "new" organization of consciousness and the "new" world of thought found in the New Testament. He claims that the divine revelation of the New Testament is given to us not in a visual but in an *auditory* mode. According to Torrance, we cannot find that the eye-witnesses tell us anything of what Jesus looked like in the gospels or epistles of the New Testament. The reason is that, he states, "the divine reality they speak of was one which they knew primarily through hearing." Torrance indicates a shift that took place in the consciousness of the disciples and apostles as follows: "It is right here that we can discern an outstanding element in the reorganisation of the consciousness of the disciples and apostles, when under the impact of divine revelation through the Word made flesh, there took place a shift away from optical forms of thought to *auditive* forms of thought arising from direct acts of cognition in hearing God."[320] Hence, he contends that we can interpret the Scriptures rightly "only in their correlation with the Word of life, the personal auditory self-revelation of God the Father embodied in him, as their enlightening, controlling and empowering centre."[321]

As Gunton has also noted, when visual patterns of perception dominate our understanding, there take place many difficulties. Even though a

319   Torrance, *Space, Time and Incarnation*, p. 26.
320   Torrance, *The Christian Doctrine of God*, p. 39.
321   Torrance, *The Christian Doctrine of God*, p. 40.

The Presence of the Triune God

famous Western saying states that "seeing is believing," seeing is not believing in the Scriptures. In the wilderness, the Israelites revolted against Yahweh in spite of their seeing his many wondrous miracles through his servant Moses. Most of the contemporaries of Jesus also did not believe in him in spite of their seeing his many wondrous miracles. Paul proclaims, "We live by faith, not by sight" (2 Cor. 5:7). In the epistle to the Hebrews, we also find the following saying: "what is seen was not made out of what was visible" (Heb. 11:3b). It is right that seeing sometimes deludes us. Yet seeing also sometimes plays an important role in our faith. Here is the first verse of the First Epistle of John: "That which was from the beginning, which we have heard, which we have seen with our eyes, which we have looked at and our hands have touched – this we proclaim concerning the Word of life" (1 Jn. 1:1). Thus, the emphasis on the hearing is not the whole story of the New Testament.

Citing Schoenberg's opera *Moses und Aron*, Richard Viladesau maintains that there are tensions between two traditions within Christianity. Moses and Aron are the representative of each tradition. In *Moses und Aron*, Aron says to Moses, "I was to speak in images, while you speak in ideas; I speak to the heart, while you speak to the mind."[322] Viladesau states that "Schoenberg insists that the genuine and beatifying encounter with God occurs when we abandon our attempts to use God, and instead submit ourselves to God's transcendence."[323] God is beyond conception and imagination. As a result, God is revealed not only in word but also in images, insofar as these mediate the enlightenment that comes from an encounter with the living God. Viladesau asserts, "the total prohibition of images is not part of the original intent of the law, but is an example of the construction of a 'hedge' around the law by the rabbis."[324] Thus, he

---

322 Richard Viladesau, *Theological Aesthetics: God in Imagination, Beauty, and Art* (New York/Oxford: Oxford University Press, 1999), p. 47f.
323 Viladesau, *Theological Aesthetics*, p. 50.
324 Viladesau, *Theological Aesthetics*, p. 53. Viladesau refers to von Rad's contention about the prohibition of images and presents an objection: "Von Rad calls the prohibition of depictions of God 'intrinsic' and 'fundamental' to the Old Testament, despite the difficulty this posed for popular religion. Not only were images of alien gods to be shunned but the God of Israel could also not be physically 'portrayed.' Aidan Nichols cautions, however, that we should not oversimplify the situation of aniconism in ancient Israel" (ibid., p. 52).

242 *Transcendence and Spatiality of the Triune Creator*

avers that "the most radical critique of idolatry flees from both image and word, to seek mystical immediacy with God."[325]

In his recent book, *Visual Faith: Art, Theology, and Worship in Dialogue*, "[c]ontrary to what our tradition may have taught us," Dyrness states, "making beautiful forms is theologically connected to our call both to listen and respond to God in prayer, praise, and sacrament."[326] He describes the abhorrence of Protestant Christians to visual arts, such as painting, sculpture, and graphic art as follows: "The spaces made for worship were not friendly to elaborate visual elements, for they were seen as distractions from true worship, which always focused on the preached Word. Creative people, then, understandably turned their imaginative energies in other directions – literature and music became special foci for Protestant creativity."[327] Dyrness acknowledges the critical importance of the spoken and written Word in Christian worship. Since the contemporary generation has been raised and nourished by images, however, he asserts that the visual imagination is inescapable: "Regardless of whether one considers this good or bad, for this generation, aesthetics counts more than epistemology."[328] Dyrness understands the Christians' experience of worship as the opportunity of responding to the gracious presence of God with the whole of their beings. The experience of worship is "an embodied experience, involving standing, kneeling, or lifting of hands; it is a deeply emotional and intellectual response of the heart to God's offer of grace in Jesus Christ by the power of the Holy Spirit."[329]

---

325 Viladesau, *Theological Aesthetics*, p. 58f. According to Viladesau, Calvin opposes "the argument (classical from the time of Gregory the Great) that images are the books of the uneducated; this view for him not only implies a class structure in religion but also assumes that pictures communicate more readily than stories" (ibid., p. 55).

326 W. Dyrness, *Visual Faith: Art, Theology, and Worship in Dialogue* (Grand Rapids: Baker Books, 2001), p. 9.

327 Dyrness, *Visual Faith*, p. 13.

328 Dyrness, *Visual Faith*, p. 20. He avers that "our children and their friends have been raised in a different world; they are often uninterested in our traditional word-centered media. Instead, they are looking for a new imaginative vision of life and reality, one they can see and feel, as well as understand [...] It is possible that we might actually win the battle of words but lose the battle of images. And losing that battle could well cost us this generation" (ibid., p. 21).

329 Dyrness, *Visual Faith*, p. 22.

The Presence of the Triune God 243

Traditionally, while the Word of God is defined as the invisible sacraments, the sacraments are defined as the visible Word of God. This visible Word of God is distinct but not separate from the audible Word of God or Holy Scripture. "In the traditional Augustinian definition," according to Muller, "a sacrament is a visible sign of an invisible grace."[330] He presents three characteristics which sacraments must properly and strictly have, as follows: "(1) they must be commanded by God *(a Deo mandata)*, (2) they must have visible or sensible elements prescribed by God, and (3) they must apply and seal by grace the promise of the gospel *(promissio evangelicae)*."[331] By this strict definition, he excludes other rites considered sacraments by Rome except baptism and the Lord's Supper. Whereas such rites as confirmation, penance, ordination, and extreme unction do not have the *mandatum Dei*, Muller states, marriage lacks the *promissio evangelicae*. Concerning the relationship between baptism and the Lord's Supper, Moltmann provides a good explanation:

> Just as baptism is the eschatological *sign of starting out*, valid once and for all, so the regular and constant fellowship at the table of the Lord is the eschatological *sign of being on the way*. If baptism is called the unique *sign of grace*, then the Lord's supper must be understood as the repeatable *sign of hope*. Baptism and the Lord's supper belong essentially together and are linked with one another in the messianic community [...] Baptism and the Lord's supper are the signs of the church life, because they are the signs of the one who is their life. They are in this way the public signs of the church's confession of faith because they show the one who leads the world into the liberty of the divine life.[332]

By baptism, the individual enters the fellowship of Christ and starts his/her pilgrimage. By the Lord's Supper, believers sustain their linkage to the community and stay within it.[333]

---

330 Richard A. Muller, *Dictionary of Latin and Greek Theological Terms: Drawn Principally from Protestant Scholastic Theology* (Grand Rapids: Baker Books, 1985), p. 267.

331 Muller, *Dictionary of Latin and Greek Theological Terms*, p. 267f.

332 J. Moltmann, *The Church in the Power of the Spirit: A Contribution to Messianic Ecclesiology* (Minneapolis: Fortress Press, 1993), p. 243.

333 Stanley Grenz designates baptism as "the seal of our identity" and the Lord's Supper as "reaffirming our identity" [S. J. Grenz, *Theology for the Community of God* (Grand Rapids: Eerdmans, 1994), pp. 520, 531].

244   *Transcendence and Spatiality of the Triune Creator*

In the discussion on the sacraments, even though the Lord's Supper in itself is fundamental to the wealth of the church, unfortunately the mode of the presence of Christ in Lord's Supper, with the mode of baptism, has become the occasion for the misery of schism and denominational conflict.[334] I support Calvin's view on this issue. But I do not stick to it because it is absolutely right but because it promotes the church unity. His view tries to avoid two extremes. Concerning Calvin's view on the presence of Christ in the Lord's Supper, Stanley Grenz aptly states as follows:

> His appeal to the Spirit facilitated Calvin in charting a middle position. He agreed with the Roman Catholic theologians and Luther that Christ's presence at the Eucharist is focused on the Communion elements. However, he denied that this entails the real presence of the Lord's physical body. Here he agreed with Zwingli that Christ in his human nature is localized in heaven. Rather than postulating a real presence, the Geneva Reformer spoke of Christ's *spiritual* presence in the elements. But this great communion with Christ is facilitated by the Spirit. The Holy Spirit unites us with the Lord across the great distance between our location on earth and his presence at God's right hand.[335]

Calvin's view is a middle position between the Roman Catholic theologians and Luther, who contended for the real bodily presence of Christ in the bread and wine, and Zwingli, who insisted that "Christ's presence is not 'in' the bread and wine at all." With Luther, according to Grenz, Calvin concluded that, "when we receive the sacrament in faith, 'we are truly made partakers of the real substance of the body and blood of Jesus Christ.'"[336] When visualizing how this occurs, however, Calvin broke

334   Moltmann, *The Church in the Power of the Spirit*, p. 244. He contends that, "The very names by which it is called are an expression of the different aspects on which emphasis has been laid and which – when they have been held with rigid absolutism – have destroyed fellowship." Because of the ecumenical reason, he prefers 'Lord's supper' *(Herrenmahl)* to 'the Mass' or 'the sacrifice of the Mass' and the Protestant expression 'the Lord's supper' *(Abendmahl, Nachtmahl)*. According to Moltmann, the name 'Lord's supper' "points to the common christological foundation of the different church traditions" (ibid.). Unlike Moltmann, Stanley Grenz writes, "In recent ecumenical discussions the designation 'Eucharist' derived from the Greek *eucharisto* ('to give thanks to') and dating to the patristic era has gained widespread use" (S. Grenz, *Theology for the Community of God*, p. 532).

335   Grenz, *Theology for the Community of God*, p. 535.

336   Grenz, *Theology for the Community of God*, p. 535. This citation of Calvin comes from *Short Treatise on the Holy Supper*, trans. J. K. Reid, in *Calvin: Theological*

The Presence of the Triune God 245

with the Middle Ages and invoked the mediatorial work of the Spirit. "Not to diminish this sacred mystery," Calvin states, "we must hold that it is accomplished by the secret and miraculous virtue of God, and that the Spirit of God is the bond of preparation, for which reason it is called spiritual."[337]

In addition to the merits to mediate the opposite positions and to give the due emphasis on the Spirit's role in the Eucharist, Calvin's view presupposes the right understanding of the concept of space. This is one of the theses of Torrance's book *Space, Time and Incarnation*. For an ecumenical purpose of the unification of Scottish and Anglican churches, he mainly attacks the Lutheran views on the incarnation and the presence of Jesus in the Eucharist. Because they stuck to the receptacle concept of space, according to Torrance, the Lutherans argued for the *kenosis* in the incarnation of the Son and the ubiquity of Jesus' body. Unlike them, Torrance holds that the Reformed theologians succeeded in rightly explaining the mode of the presence of Jesus in the Eucharist, applying the relative and dynamic concept of space which the Nicene fathers had adopted to explain the incarnation of the Son. Behind the dispute between the Lutheran and the Reformed theologians on the presence of Jesus in the Eucharist, there are different understandings of the ascension of the resurrected Christ, which result from certain conceptualization of space. Due to their receptacle notion of space, the Lutheran theologians had to think of the incarnation as "the self-emptying of Christ into the receptacle of a human body."[338] Likewise, they asserted that the resurrected Christ could not ascend to the right hand of the Father spatially in behalf of the bodily presence of Jesus in the Eucharist. Instead of that, the Lutherans assumed a view of the ubiquity of the body of Christ. Against the Lutheran views, Torrance states, "As the Incarnation meant the entry of the Son into space and time without the loss of God's transcendence over space and time, so the Ascension meant the transcendence of the Son over space and time without the loss of His incarnational involvement in space and time."[339] Torrance understands the

---

Treatise, volume 22 of the *Library of Christian Classics* (London: SCM, 1954), p. 166.

337 John Calvin, *Short Treatise on the Holy Supper*, p. 166, as cited by S. Grenz, *Theology for the Community of God*, p. 535.

338 Torrance, *Space, Time and Incarnation*, p. 35.

339 Torrance, *Space, Time and Incarnation*, p. 31.

246 *Transcendence and Spatiality of the Triune Creator*

saying that Christ is "in heaven" as follows: In the incarnation, the Son was made flesh without ceasing to be transcendent God. In the ascension, Christ ascended above space and time without ceasing to be human being. Thus, concerning the presence of Jesus in the Eucharist, Torrance alludes, "In the incarnation we have the meeting of man and God in man's place, but in the ascension we have the meeting of man and God in God's place, but through the Spirit these are not separated from one another (they were not spatially related in any case), and man's place on earth and in the space-time of this world is not abrogated, even though he meets with God in God's place."[340] Unlike the Lutheran, then, Torrance opts for the spiritual presence of Jesus in the Lord's Supper.

According to Douglas Farrow, Luther contended that God has and knows various ways to be present at a certain place, and that the presence of Christ must be considered in at least three kinds of presence.[341] The first is the local, or "circumscriptive" mode of presence. It denotes the mode of presence of physical, finite things. In this mode, "Christ used to be present on earth, and can still be if he so chooses, as will be the case on the last day."[342] The second is the "definitive" mode of presence. It is the presence of finite spiritual beings. They cannot be circumscribed or assigned to a spatially defined *locus* because of the immaterial or nonphysical nature of the beings. But their locus, nevertheless, is limited or defined *(definitiva)* by the finitude of the beings in their power and operation. For Luther, "This is the mode of his presence in the eucharistic bread, and it does not compromise the specificity of his resurrected humanity, the powers of which we too will some day share."[343] The third is the "repletive" mode of presence. It is "the presence of a spiritual being that fills a place without being contained or defined in any

---

340 Torrance, *Space, Time and Resurrection*, p. 129f.
341 Douglas Farrow, "Between the Rock and a Hard Place: In Support of (something like) a Reformed View of the Eucharist," *International Journal of Systematic Theology* 3/2 (July 2001), p.174, note 20. Similar to Luther, Muller presents five kinds of presence: (1) *Praesentia localis sive corporalis*, local or bodily presence, or *praesentia circumscriptiva*, circumscriptive presence. (2) *Praesentia spiritualis sive virtualis*, spiritual or virtual presence. (3) *praesentia illocalis sive definitiva*, illocal or definitive presence. (4) *praesentia repletiva*, repletive presence. (5) praesentia temporalis, temporal presence (R. A. Muller, *Dictionary of Latin and Greek Theological Terms*, p. 239f).
342 Farrow, "Between the Rock and a Hard Place," p. 174, note 20.
343 Farrow, "Between the Rock and a Hard Place," p. 174, note 20.

The Presence of the Triune God

way by that place."[344] This mode belongs to God alone but also to Jesus Christ. Douglas Farrow, however, concludes that all this is unconvincing. He presents five reasons why he does not agree with Luther's view. Among them, the last two are pertinent to the present study. Following Torrance, Farrow opposes the Lutheran idea of the circumscriptive mode of presence, because this notion depends upon a false receptacle, or container, notion of space. Then, finally, Farrow claims, the eschatological nature of the Eucharist is suppressed in Luther's view.[345]

Torrance asserts that the difficulties of the Lutherans are due to their concept of space. He contends that Lutheran and Newtonian thought shared with the old receptacle notion of space. The infinite receptacle notion of space is intimately related to the absolute space concept of Newton. However, according to Torrance, Reformed and Anglican theology stood much closer to the relational view of space advocated by Leibniz, who attacked the doctrine of absolute and uniform space and time. Concerning Leibniz, D. Bertoloni Meli writes,

> In his refutation of absolute space and time Leibniz would argue that if space and time were something absolutely real and uniform, God would lack a sufficient reason for arranging the bodies in one way rather than another, changing for example east into west, or for creating the universe at one particular instant rather than another, such as a year sooner [...] However, according to Leibniz's refined argument against Clarke, if one rejects absolute space and time, all the different cases would become indistinguishable and therefore they would be identical. Thus God would not have to make a choice lacking a sufficient reason.[346]

---

344 Muller, *Dictionary of Latin and Greek Theological Terms*, p. 240. Muller's second mode, "spiritual or virtual presence," includes Luther's second and third modes of presence. "Spiritual or virtual presence" is a term applying to "any spiritual being which, since it is immaterial, does not occupy space, but rather manifests its presence by a power *(virtus)* of operation" (ibid., p. 239).

345 Farrow, "Between the Rock and a Hard Place," p. 174, note 20. The first three reasons Farrow suggests for his disputing Luther's view are as follows: "First, because the repletive mode does indeed confuse the nature, a fact to which the complete absence of any pneumatological appeal bears quiet testimony. Second, because it is illegitimate to argue, as Luther does, from the repletive mode to the diffinitive mode. Third, because the diffinitive mode cannot be equated with the eucharistic mode without unwanted implications" (ibid.).

346 D. Bertoloni Meli, "Caroline, Leibniz, and Clarke," *Journal of the History of Ideas* (1999), p. 483.

248 *Transcendence and Spatiality of the Triune Creator*

Even though we easily recognize that Leibniz was consequently right in that he contended for a thoroughly relational view of space, the context of the controversy between Leibniz and Clarke was so complex that it needs a more careful examination than we are able to give it here.[347]

The Lutheran theologian Pannenberg observes that "the contrast between Leibniz and Newton is repeated in the relationship of Whitehead's philosophy of nature to the metaphysical conceptions that historically lie at the basis of the field concept."[348] This contrast is between the conception of space as a condition for the reality of bodies and movements (Newton), on the one hand, and the reduction by Leibniz of space and time to invisible, punctiliar entities, or monads, on the other hand. Pannenberg interprets Peacocke's conception of the relationship of God to the world in analogy to the relationship of soul and body as close to the Newtonian conception. Pannenberg maintains that, like Peacocke, "Newton also wanted to be neither a dualist nor a pantheist. But he had more exact conceptions of the relationship of God to the world."[349] In this sense, Pannenberg seems to support the view of Newton over that of Leibniz. Pannenberg, however, does not hold a typically Lutheran view on the Eucharist.

347 Meli points out that Leibniz was a Lutheran and his life-long concern was church union. According to Meli, Leibniz defended the Lutheran idea of a "concomitance" of the bread and wine on the one hand and the body and blood of Christ on the other. It is the idea that both substances are received at the same time in the Eucharist. Leibniz's doctrine of the Eucharist requires a *real* and *substantial* presence of the body of Christ. Meli asserts that "In Leibniz's system, unlike Newton's, substance is endowed with activity and passivity, not with extension, which appears only at the level of phenomena or physically [...] Beginning very early in his career Leibniz tried to present such views, namely, the removal of the notion of substance, as a way to reconcile different doctrines of the eucharist, such as the Calvinist, the Lutheran, and initially also the Catholic" (Meli, "Caroline, Leibniz, and Clarke," p. 479). Thus, Meli states that Leibniz's views on substance, gravity, space, and time evolved over several decades hand in hand with his theological concerns and were an eminently suitable ground for church reunion (ibid., p. 484). This description of Meli is different from Torrance's arguments that Lutheran theology and the Newtonian system shared the absolute concept of space and that Reformed and Anglican theology stood much closer to the relational concept of space. In this controversy between Leibniz and Clarke, Meli indicates the important role of Caroline, Princess of Wales (ibid.). Pannenberg also refers to this Princess' existence in this controversy [Pannenberg, *Toward a Theology of Nature*, p. 59].

348 Pannenberg, *Toward a Theology of Nature*, p. 58.

349 Pannenberg, *Toward a Theology of Nature*, p. 59.

The Presence of the Triune God 249

Pannenberg wants to revise "traditional ideas of the heavenly corporeality of the risen Lord becoming present in the bread and wine." Like Stanley Grenz, Pannenberg takes Calvin's view on the eucharistic presence of the Lord, as a middle stance between Zwingli and Luther. Pannenberg argues that,

> On the one hand he [Calvin] rejected a presence of Christ in the elements, treating these only as *signs* of Christ's body and blood. On the other hand he taught a spiritual eucharistic presence of Christ, even according to his humanity, which we can grasp by faith. In this regard the hidden work of the Holy Spirit was decisive for Calvin. The transfigured flesh of Christ does not enter into us, but by the Spirit his life is imparted to our souls.[350]

Unlike orthodox Lutherans, Pannenberg emphasizes "the presence of Christ's whole person, not especially the bodily aspect of its reality."[351] In this regard, he proposes Calvin's view as the overcoming the inner Reformation differences in eucharistic teaching.[352] Thus, Pannenberg holds, "We are not to think of Christ's presence at the Supper as a direct descent of the risen Lord in the form of bread and wine as at the incarnation, notwithstanding the expressiveness of that comparison within its limits. Instead, we are to think of it in terms of recollection of the earthly story of Jesus and his passion."[353] Like Calvin, Pannenberg aptly emphasizes the role of the Spirit in the Eucharist. But, like Luther, Pannenberg opts for the presence of Jesus to his community in the elements of bread and wine. Unlike Luther, however, as "the medium or locus of his presence to believers," Pannenberg does not appeal to the Lutheran view of

350 Pannenberg, *Systematic Theology* 3, p. 312. Concerning the Lutheran view, Pannenberg writes, "In contrast, the Lutherans in the Formula of Concord insisted on Christ's bodily presence at the Supper because we would have to call communion with Christ in the eucharistic bread a fellowship not with his body but with his spirit, power, and benefits if the body of Christ were present and taken only according to its force and efficacy. Rejected here was the Reformed view that after his ascension Christ is tied spatially to heaven according to his humanity, so that he cannot be present bodily on earth in the sacrament" (ibid., p. 313).
351 Pannenberg, *Systematic Theology* 3, p. 313.
352 Pannenberg, *Systematic Theology* 3, p. 314, states, "It says indeed that Christ imparts himself 'with' the distributing of bread and wine, but it does not define more precisely the relation to these 'elements' at the Supper nor say anything at all about the 'consecration' of bread and wine. Thus the basis of the agreement is more that of Calvin than of Luther."
353 Pannenberg, *Systematic Theology* 3, p. 315.

250   *Transcendence and Spatiality of the Triune Creator*

the ubiquity of Jesus' body but the "anamnesis as recollection of Jesus' institution of the Supper at the parting from the disciples."[354]

In his book, *Eucharist and Eschatology*, Geoffrey Wainwright traces the imagery of the Eucharist as the anticipatory meal with eschatological implications.[355] The cry "Come, Lord Jesus" (*maranatha*, 1 Cor. 16:22; cf. Rev. 22:20) is not only a request for his coming for table fellowship in anticipation of God's coming kingdom (*Didache* 10.6) but also a request for the eschatological coming of the ascended Lord to consummate his kingdom.[356] Thus, the Eucharist has this eschatological dimension. Farrow contends, "The eucharistic site is an *eschatological* site, situated between the first and the final parousia."[357] Indicating the difficulty of the idea of a spatial presence that ignores time, Torrance states, "If we posit any kind of spatial relation without extension in time we make it impossible to discern any difference between the real presence of Christ in the days of His flesh, in the Eucharist, and at the Last Day; but that is equivalent to making the historical foundation of faith irrelevant."[358] The presence of Christ in the Eucharist is not the same as the *parousia* at the Last Day. The Eucharistic presence of the Lord is an anticipation of his bodily second coming at the eschaton. Then he will accomplish the new creation that he already started during his first coming.

4.5 New Creation[359]

Moltmann broadens the doctrine of creation to embrace not only creation in the beginning, but also creation in history, and the creation of the End-time: *creatio originalis – creatio continua – creatio nova*. For him, "'creation' is the term for God's initial creation, his historical creation,

---

354   Pannenberg, *Systematic Theology* 3, p. 323. He refers to the Luther's doctrine of ubiquity just one time in his footnote (ibid., p. 313, note 682).
355   Geoffrey Wainwright, *Eucharist and Eschatology* (New York: Oxford University Press, 1981).
356   Cf. Pannenberg, *Systematic Theology* 3, p. 320. This cry, according to Pannenberg, "can be construed as a petition but also as a proclamation of the Lord's presence in the Supper."
357   Farrow, "Between the Rock and a Hard Place," p. 180.
358   Torrance, *Space, Time and Incarnation*, p. 35.
359   Since eschatology is a time-related term, I use this term "new creation" to highlight the spatial aspect.

The Presence of the Triune God 251

and his perfected creation."[360] If the doctrine of creation includes eschatology, he contends, "eschatology is nothing other than faith in the Creator with its eyes turned towards the future."[361]

In order to escape a gnostic doctrine of redemption, "Christian eschatology must be broadened out into cosmic eschatology,"[362] argues Moltmann. Cosmic eschatology as the true Christian eschatology is to teach not a redemption *from* the world but the redemption *of* the world, and not a deliverance of the soul from the body but the redemption of the body. He maintains that this eschatology is required for God's sake, not for the sake of some universalism or other.

In this argument, Moltmann tends toward a version of universalism. Yet, he claims to overcome two opposite views: "universalism" versus "a double outcome of Judgment," or particularism. Nevertheless, his alternate to these two views is actually a kind of universalism. Moltmann asserts as follows:

> Paul and John talk about 'being lost' only in the present tense, never in the future. So unbelievers are 'given up for lost' temporally and for the End-time, but not to all eternity. This being so, we can conclude with Walter Michaelis that what is said about judgment, damnation and 'everlasting death' is aeonic, and belongs to the End-time; it is not meant in an 'eternal' sense. For eschatologically, against the ho-

360 Moltmann, *God in Creation*, p. 55.
361 Moltmann, *God in Creation*, p. 93. Similar to Moltmann, in the foreword of *Systematic Theology*, vol. 2, Pannenberg contends, "In a certain sense creation will be complete only with the eschatological consummation of the world" (Pannenberg, *Systematic Theology* vol. 2, p. xvi). According to him, "Creation and eschatology belong together." For "it is only in the eschatological consummation that the destiny of the creature, especially the human creature, will come to fulfillment." However, they are not directly identical, at least from the creature's standpoint, but form a unity only from the view of the divine act of creation (ibid., p. 139). Concerning the relationship between creation and eschatology, Pannenberg asserts, "If the eschatological future of God in the coming of his kingdom is the standpoint from which to understand the world as a whole, the view of its beginning cannot be unaffected. This beginning loses its function as an unalterably valid basis of unity in the whole process. It is now merely the beginning of that which will achieve its full form and true individuality only at the end. Only in the light of the eschatological consummation can we of the world understand the meaning of its beginning" (ibid., p. 146). Therefore, Pannenberg's concern about the end comes from his concern about the whole. At the end of the history of this world we can get to truth as a whole.
362 Moltmann, *The Coming of God*, p. 259.

252   *Transcendence and Spatiality of the Triune Creator*

rizon of the ultimate, it is penultimate. The ultimate, the last thing is: 'Behold, I make *all things* new' (Rev. 21.5).[363]

This position is the logically necessary result of his ecological doctrine of creation. *Creatio originalis*, creation in the beginning started with nature and ended with the human being. *Creatio nova*, the new creation reverses this order: it starts with the liberation of the human being and ends with the redemption of nature. Thus Moltmann states: The new creation at the eschaton will not only manifest the liberty of the children of God. This eschatological creation will also bring "the deification of the cosmos" through the unhindered participation of all created beings in the livingness of God.[364] From Moltmann's perspective of the redemption of the whole nature, we would conclude that it is impossible to exclude unbelievers from eternal salvation.

According to Moltmann, the confidence that nothing will be lost is not the optimistic dream of a purified humanity. Everything will be brought back again and gathered into the eternal kingdom of God. As the reason for why we can have these confidences, he presents the cross of Jesus and his descent into hell. He asserts the following: *"The true Christian foundation for the hope of universal salvation is the theology of the cross, and the realistic consequence of the theology of the cross can only be the restoration of all things."*[365] He also holds the Arminian view that Christ 'per suam passionem destruxit totaliter infernum' (through his passion destroyed hell totally), even though this opinion was condemned by the patristic church.[366] "To make Christ's death on the cross the foundation for universal salvation and 'the restoration of all things,'" Moltmann thus maintains, "is to surmount the old dispute between the universal theology of grace and the particularist theology of faith."[367]

The distinction between the universal theology of grace and the particularist theology of faith, however, is not appropriate. Against Arminians, Calvinists have emphasized the absolute sovereignty of God. Al-

---

363   Moltmann, *The Coming of God*, p. 242.
364   Moltmann, *The Coming of God*, p. 92.
365   Moltmann, *The Coming of God*, p. 251.
366   Moltmann, *The Coming of God*, p. 374, note 245. See also J. Moltmann, "The Logic of Hell," *God's Being All in All*, pp. 43–47.
367   Moltmann, *The Coming of God*, p. 254.

The Presence of the Triune God

253

though they insist on the irresistibility of grace, Calvinists argue for limited salvation. Contrarily, Arminians have a tendency toward universalism, even though they have emphasized the need for human response. Thus, to draw a dichotomy between the universal theology of grace and the particularist theology of faith does not accurately reflect the historical data.[368]

In his book, *God in Creation*, in which he sets forth the ecological concept of creation, Moltmann states the following: "Faith in God the Creator cannot be reconciled with the apocalyptic expectation of a total *annihilatio mundi*. What accords with this faith is the expectation and active anticipation of the *transformatio mundi*."[369] But Moltmann goes even further. Concerning ideas about the *consummatio mundi*, the consummation of the world, he proposes three mainline views: 1) The total *annihilation* of the world according to orthodox Lutheranism; 2) The total *transformation* of the world according to patristic and Calvinistic tradition; and 3) The world's glorious *deification*, the view of Orthodox theology.[370] Concerning these options Moltmann suggests a criterion for evaluating which one is right: "Christian eschatology has its foundation in the experience of Christ's death and resurrection. Cosmic eschatology also belongs within the framework of this remembered hope for Christ: the death and raising of the universe are the prelude to the expected new creation of all things and 'the new heaven and the new earth.'"[371] Moltmann presents his comments on each position as follows:

368 Cf. Paul Helm, *Calvin and the Calvinists* (Carlisle, PA: The Banner of Truth Trust, 1982), refutes Kendall's contention that, whereas Calvin contended for general atonement and salvation by grace through faith, his successors, the Puritans, changed Calvin's views and supported a limited atonement and salvation through good endeavours (p. 9). Helm states that there is no evidence to support Kendall's idea of sharp doctrinal difference between Calvin and the Puritans. According to Helm, both Calvin and the Puritans emphasized limited atonement and faith as trust in the divine promises (p. 71f).

369 Moltmann, *God in Creation*, p. 93.

370 Moltmann, *The Coming of God*, p. 267f. Concerning the "deification" of the world in the Orthodox churches, Moltmann says, "'Deification' [...] does not mean that human beings are transformed into gods. It means that they partake of the characteristics and rights of the divine nature through their community with Christ, the God-human being. The divine characteristics of non-transience and immortality therefore become benefits of salvation for human beings" (ibid., p. 272).

371 Moltmann, *The Coming of God*, p. 261.

254 *Transcendence and Spatiality of the Triune Creator*

> The Lutheran doctrine of the annihilation of the world seems to have as its premise
> a one-sided theology of the cross. The Orthodox doctrine of deification, on the
> other hand, corresponds to a one-sided theology of the resurrection. The Calvinist
> theory of transformation could be the mediation between perspectives directed to-
> wards the end of 'this world,' and the genesis of a 'new world' that will accord
> with God and thus be deified.[372]

But the Calvinist view of transformation, according to Moltmann, has never in its history been able to attain either to "the depths of the Lutheran theology of the cross" or to "the heights of the Orthodox theology of deification." Thus he holds that "the *reductio in nihilum*, the reduction to nothingness, and the *elevatio ad Deum*, the elevation to God, belong together and are mutually complementary."[373]

Moltmann defines the new creation in view of a new divine presence within it. So the Creator no longer remains over against his creation. God the Creator dwells in it, and finds in it his rest. This divine presence within it makes of the new creation "a sacramental world." The new creation is interpenetrated by divine presence and participates in the inexhaustible fullness of God's life. Thus, he argues as follows: "The indwelling of God calls into being a kind of cosmic *perichoresis* of divine and cosmic attributes. In that new aeon a mutual perichoresis between eternity and time also comes into existence, so that on the one hand we can talk about 'eternal time' and on the other about 'eternity filled with time.'"[374]

Concerning the new creation as "the work of making the earth a suitable vehicle for God's glory," Dyrness states, "though we know little about the details of this event, we do know that it will be an embodied event."[375] In Rev 21:2, John describes the new creation as "a bride adorned for her husband" (NIV). According to Dyrness, this expression of the bride is used of the body of Christ elsewhere in the New Testament (see 2 Cor 11:2). Thus, Dyrness contends that "the creation itself will share in the feast of the marriage supper of the lamb, which is prepared for the body of Christ." Citing another verse of the Revelation of John, "See, the home of God is among mortals" (21:3), Dyrness indicates that "the direction of movement is not upward toward heaven and

372 Moltmann, *The Coming of God*, p. 274.
373 Moltmann, *The Coming of God*, p. 274.
374 Moltmann, *The Coming of God*, p. 295.
375 Dyrness, *The Earth is God's*, p. 56.

The Presence of the Triune God 255

the transcendent 'place' of God but downward toward the human creation. It is as though the incarnation itself foreshadows this final created revelation of God's work, where God's very dwelling is with the creature."[376]

The glorification of God is the final article in dogmatics, especially the dogmatics of Calvinist orthodoxy. It is the ultimate purpose of creation. Thus the supreme goal of human beings is "to glorify God and to enjoy him forever." Moltmann maintains that "This glorifying of God in the world embraces the *salvation* and eternal life of human beings, the *deliverance* of all created things, and the *peace* of the new creation."[377] For Moltmann, "To 'glorify' God means to love God for his own sake, and to enjoy God as he is in himself."[378] The glorification of God has no purpose and utility – if it had, God would not be glorified for his own sake, contends Moltmann. Thus he relates the glorification of God to the child's self-forgetting delight in its game. He argues: "Free human self-expression is an echo of the Creator's good pleasure in the creations of his love. Consequently the simplest glorification of God is the demonstrative joy in their existence of those he has created."[379] In the discussion about the glorification of God, however, instead of the significance of the glorification of God for all his creatures, including human beings, Moltmann turns the question upside down and asks about the meaning *for God himself* of his glorification by all his creatures. He quotes rhetorically the following (perhaps overly economic) question from the book *Theodramatik*, by Hans Urs von Balthasar: "What does God get from the world?"[380]

Concerning God's glory Moltmann presents three theses which have been put forward in the course of the history of theology and discusses their consequences. He writes:

---

376 Dyrness, *The Earth is God's*, p. 57.
377 Moltmann, *The Coming of God*, p. xvi.
378 Moltmann, *The Coming of God*, p. 323.
379 Moltmann, *The Coming of God*, p. 323f.
380 Moltmann, *The Coming of God*, p. 324. Here Moltmann lists the following questions: "Is the world a matter of indifference for him, because he suffices for himself? Does he need it, in order to complete himself? Does it perhaps give him pleasure, because he rejoices in the echo of those he has created? Is there also a kind of divine eschatology in the glorification of God, so that in his glorification God arrives at his goal, and in his goal arrives at himself?" (ibid.)

256   *Transcendence and Spatiality of the Triune Creator*

> Does glory consist 1. in the self-glorification of God? 2. in the self-realization of God? 3. in the interactions between divine and human activity? True glory is: 4. The fullness of God and the feast of eternal joy.[381]

God is not glorified only by God himself but can also be glorified through an interaction between God and the world, between God and human beings, or between the trinitarian divine Persons. Moltmann proposes thinking about the co-workings which lead to glorification in trinitarian terms. Over against the tradition of God's self-glorification, Moltmann argues, "in the beginning God in himself – at the end God all in all. In this divine eschatology God acquires through history his eternal kingdom, in which he arrives at his rest in all things, and in which all things will live eternally in him."[382] Over against Hegel's idea of God's self-realization, Moltmann argues, "the history of Christ's self-emptying and glorification is not to be thought of modalistically; it is seen in trinitarian terms, as the co-workings of the Father, the Son and the Spirit." Without thinking in trinitarian terms about God's self-realization, our thinking is led either to a dialectical pantheism or to an apotheosis of the world. If we think of it trinitarianly, Moltmann asserts, "all created beings are drawn into the mutual relationships of the divine life, and into the wide space of the God who is sociality."[383] Over against the interactions between divine and human activity, states Moltmann, "faith and the discipleship of Christ awaken a strong assurance that through Christ the name of God is sanctified, that in him God's will is done, and that with him, therefore, God's kingdom will come."[384] He finds the genesis of "the trinitarian ideas about the co-workings of the Son and the Father and the Spirit of truth" in close spiritual proximity to Jewish Shekinah theology.

Concerning the fourth thesis, Moltmann maintains that "The approach by way of *the fulness of God* (Eph. 3.19) which 'dwells bodily' in Christ leads us beyond the traditional ideas about the self-glorifying *will* of God and the self-realizing *nature* of God and makes the *interplay* of all blessing and praising, singing, dancing and rejoicing creatures in the commu-

---

381 Moltmann, *The Coming of God*, p. 324.
382 Moltmann, *The Coming of God*, p. 335.
383 Moltmann, *The Coming of God*, p. 335f.
384 Moltmann, *The Coming of God*, p. 336.

The Presence of the Triune God

257

nity of God more comprehensible."[385] The fulness of God and the rejoicing of all created beings prepare the feast of eternal joy. Moltmann maintains, "If we could talk only about God's nature and his will, we should not do justice to his plenitude."[386] Moltmann acknowledges that a human analogy is bound to be inappropriate. Thus in thinking of the fulness of God, he refers to "the inexhaustibly rich *fantasy of God*, meaning by that his creative imagination." Finally Moltmann goes beyond the view of creation as the free decision of God's will and the view of creation as God's self-realization:

> If creation is transfigured and glorified, as we have shown, then creation is not just the free decision of God's will; nor is it an outcome of his self-realization. It is like a great song or a splendid poem or a wonderful dance of his fantasy, for the communication of his divine plenitude. The laughter of the universe is God's delight. It is the universal Easter laughter.[387]

He thus adopts an aesthetic concept to describe the final glorification of God. It reminds me of a saying of Moltmann in *God in Creation*: "The eternal perichoresis of the Trinity might also be described as an eternal round danced by the triune God, a dance out of which the rhythms of created beings who interpenetrate one another correspondingly rise like an echo."[388]

This eschatological vision of Moltmann is at some point not acceptable. Though I admit that his vision is grand and magnificent, I am not sure that the picture which Moltmann wants to present by means of the cosmic perichoresis of the infinite and the finite is biblical. In the final state, this creation will become a suitable vehicle for God's glory. Moltmann's vision implies the divinization of the world. However, it also seems to entail some restrictions on God. Concerning the details of new creation at the eschaton, we know little, but we know it is an embodied event, Dyrness states. In this sense, Moltmann proposes too

---

385 Moltmann, *The Coming of God*, p. 336.
386 Moltmann, *The Coming of God*, p. 338.
387 Moltmann, *The Coming of God*, p. 338f.
388 Moltmann, *God in Creation*, p. 307. Interestingly, Moltmann finds the ground of the Reformed doctrine of the predestination not in theology but in aesthetics. He contends, "Antithesis in art make for symmetry [...] The aesthetic of juxtaposition is the inner motive for the doctrine of double predestination" (Moltmann, *The Coming of God*, p. 247).

258   *Transcendence and Spatiality of the Triune Creator*

much. He goes too far. Moltmann goes beyond God's will or nature. He appeals to God's fantasy. Even though his picture of the new creation is highly suggestive, it does not totally have a biblical basis. His vision goes beyond the limits of the Scriptures and offers the fantasy of Moltmann himself.

5. Conclusion

God is not only the creator of the world. But he is also the preserver. God is transcendent over the world of space and time. He is immanent and present in the world of space and time. Dyrness's categories of relationship, agency, and embodiment are grounded in the presence of the triune God. God is not only internally relational but also externally has relationship with the world of creation. God interacts with the world. Even though he cannot be considered as one of the creaturely causes, God can act as an agent. The world is not a body of God. But God's glory is embodied through material things, including the human body.

In this chapter, I have asserted that theology needs to find a way to speak of divine spatiality, if it is to speak adequately about the presence of the triune God. By virtue of being present in the world of space and time, God is spatial, or has his own spatiality. According to three categories of Dyrness, I have presented three meanings, or dimensions, of God's spatiality. The triune God is not an isolated being but a relational being and has a kind of relationality. For this relationship, God needs his own spatiality. With the term "God's spatiality," thus, I have emphasized that the triune God exists in a relation or has a relationality. God's relationality is the first meaning of divine spatiality.

We may also now speak of God's agency, or activity in the world, as a second meaning of spatiality. The development of modern natural sciences narrowed down the scope of God's action and finally exiled him out of the universe. But the universe as a closed system is now questioned by scientists themselves. This world is not closed but open and the events of this world are not predictable but unpredictable. Hence there are spaces, rooms, or "gaps" for God's action in this world. Chance in biological evolution, indeterminacy in quantum level, and chaos in

The Presence of the Triune God 259

macro level are held to signal conceptions of a new openness in the shape of reality and gaps for God's action. The God who is active in this world or the God as an agent has space, room, or gaps for his own activities within the world or universe. This is the second meaning of God's spatiality.

Finally, in speaking of God's spatiality, we may speak of embodiment. In the East, this difficult discussion has focused on God's presence through the incarnation; in the West, the same issues of spatiality have been discussed with respect to the Eucharist. Christian Eschatology is yet the third doctrine with implications for refining a meaning of embodiment, or presence of God. In this discussion, I have assumed that God is not a bodily being. Nor does he have his own body. I have also rejected the view of the world as the body of God. In this world, however, God has embodied himself in various ways. We may speak of creation as the theater for God's glory. Furthermore, in the incarnation of the Son, God embodied himself in space and time. Ecclesiologically, we see that the church of both the Old and New Testaments is the earthly-historical form of God's own existence. The Eucharist is the visual and material means of God's invisible grace. Christ is in some "real" sense present in the Eucharist. The new creation is and will be the work of making the creation a suitable vehicle for God's glory. It will be an embodied event. In some sense, it has already started. God uses space, time and bodies as vehicles of his grace. By means of them, God expresses his goodness and love for his creatures. Thus, we stop short of the idea of the literal embodiment of God in the world but acknowledge a kind of divine embodiment. In these processes of embodiment, God uses material beings, including the human body, as vehicles for his embodiment.

Conclusion

Summary and Case for Classical Theism

In this book, I have treated the issue of God's spatiality in relation to the divine transcendence. While the divine spatiality is intimately related to the presence or immanence of God in the world, we must not forget to heed God's transcendence first of all. Otherwise, we fall prey to panentheism. Many theologians acknowledge God's temporality. They, however, have reluctance to admit God's spatiality. If the relational or relative concept of space is adopted, I think we can speak of God's spatiality. Not by the absolute, but by the relative space, the co-presence of the divine and the human in Jesus is possible. God is figuratively our space. God uses space as the medium of his creative presence at the finite places. God can be in the limited space of this world without his transcendence being abandoned.

There are two rival concepts of space: receptacle idea of space and relational idea of space. The receptacle concept of space has played a troublesome role in the history of the Christian theology. Patristic Fathers accepted the relational concept of space in order to develop their doctrines of creation, incarnation, and God's interaction with the world. The infinite receptacle idea of space is intimately related to the absolute space of Newton. In some sense, this concept of space was necessary for the development of modern natural sciences. But the relativity theory of Einstein disputed this notion of absolute space and time. Space and time are not an absolute and independent reality from this world. They are neither the sense organ of God nor *a priori* forms of intuition. Thus, Augustine's view that space and time were created with the world can be explained as an anticipation of the relativity theory.

Torrance contends that the concept of space of the Patristic Fathers is the relational or relative one, not the receptacle one. He identifies the relational space concept of the Patristic fathers with that of Einstein. Furthermore, he regards the space concept of Reformed and Anglican theologians in the same light with this relational concept of space. Tor-

262 *Transcendence and Spatiality of the Triune Creator*

rance states that Lutheran theology takes the Newtonian concept of absolute space. Despite its interesting and constructive points, this contention is just an ex post facto. As a possibility of the place of God, Torrance refers to *perichoresis*. But he does not fully develop this idea in relation to the discussion of divine spatiality.

Moltmann classifies the concepts of space into three: geometrical, ecological, and perichoretic concept of space. In his doctrine of creation, he proposes the ecological concept of space. But, in his eschatology, Moltmann is not satisfied with this idea of space but deepens his thought on space and presents the perichoretic concept of space. This concept of space is not definitely a receptacle one. This perichoretic concept of space means mutual interpenetration or mutual indwelling. As the perichoretic concept of space, he wants to present a kind of the divine space. In this sense, he goes one step further than Torrance. However, unlike Einstein and Torrance, Moltmann does not rightly distinguish the receptacle idea of space from the relational concept of space. He identifies the omnipresence of God with the absolute space and the relationships and movements in the created world with the relative space. The absolute space is not to be identified with the omnipresence of God. With the aid of the concept of the relative space, we can express the divine omnipresence. Thus, this thought is a kind of confusion.

In some sense, discussion of God's spatiality seems to contradict to that of divine transcendence. In fact, however, God's transcendence does not necessarily exclude divine spatiality. Some contemporary theologians cast away the divine transcendence in a spatial sense and adopt that attribute in a temporal meaning (Theology of Hope) or reduce it to an immanent meaning (Tillich). In those cases, some troublesome problems take place. The crucial problem is that their view on divine transcendence is not complete. In that sense, therefore, we cannot affirm the doctrine of divine transcendence effectively without some spatial categories. Even though the discussion on God's spatiality seems to focus mainly on God's immanence or presence in the world, my proposal is that the argument on the divine transcendence guides and limits the scope of the discussion on God's spatiality.

Due to the environmental threat of pagan pantheism, the traditional Christian doctrine of God has emphasized divine transcendence. The distinction between the Creator and creature is more important than anything. Thus, even though classical theism does not abandon divine im-

Conclusion 263

manence or presence in this world of space and time, it cannot give due significance to it. In comparison to this leaning to the divine transcendence of traditional theism, our times are the age of immanence. The dominant concern of the modern Christian theology is on the divine immanence. In a reaction to classical theism's undue emphasis on divine transcendence, many contemporary theologians take an option of panentheism as an alternative to classical theism. But this tendency is another extreme. We should hold both the creative tension of the divine transcendence and immanence. The relation of divine transcendence and immanence is not an either-or relation. Therefore, since our times are the age of immanence, I think, there is a crisis of transcendence. We should avoid "immanent transcendence." "Absolute transcendence," however, does not fit to the biblical data. If the absolute transcendence is the only option, there is no revelation at all and we do not know God at all.

Panentheism is spotlighted as a useful means to hold both the transcendence and immanence effectively. In opposition to theism and pantheism, panentheists argue for the propriety of their position. Panentheism, however, does not distinguish God from the world. After all, it slides into pantheism. Panentheism does not confirm the divine transcendence rightly. On the other hand, deism, which appeared along with the development of natural sciences in modern times, gives emphasis on the divine transcendence one-sidedly. It never mentions God's providence after the creation. Deism drives the Creator God out of the process of this world. It denies divine immanence. Panentheists usually identify classical theism with deism, since both of them share the emphasis on the divine transcendence. But this contention is not valid. Unlike deism, classical theism admits divine immanence. Otherwise, classical theism could not admit the incarnation of the Son. It contends for the continuous interaction of the Creator God within the world.

God is not only transcendent over the world but also immanent within it. God is present and interacts in this world. God has the positive relationship with the world of space and time. Thus, the divine transcendence does not necessarily exclude divine spatiality as well as temporality. But we caution that we should not read our time and space into God. God is the Creator of time and space. In this sense, like Gunton, I agree with Augustine's view that God is not a temporal being like any other creatures. At the same time, however, I do not concede Augustine's view

264   *Transcendence and Spatiality of the Triune Creator*

that God is timeless or spaceless. God has a kind of temporality and spatiality.

I examined the theological thoughts related to divine spatiality of three theologians: Torrance, Pannenberg, and Moltmann. In comparison to the latter two, Torrance does not fully develop God's spatiality, even though in some sense he confirms God's temporality. In comparison to the first, Pannenberg and Moltmann affirmatively hold divine spatiality as well as temporality. For them, God's spatiality does not contradict his transcendence. Both of them try to explain God's intimate involvement within the world of space and time. But their views on the Spirit as a field (Pannenberg) and the divine *zimzum* as creation (Moltmann) betray the tendency of panentheism. Even though they contend that panentheism holds together both the transcendence and the immanence of God in relation to the world, panentheists do fail to ensure divine transcendence. Therefore, although we can acknowledge that they make a considerable contribution to expounding divine immanence within the universe, we can not recognize that they succeed in balancing between God's transcendence and immanence.

Torrance presents valuable insights for the discussion of divine spatiality. He proposes the idea of the multi-levelled structure of human knowledge. By means of his thought on the redemption of space, he tries to differentiate between different kinds of space. He states divine temporality. But he does not fully develop these thoughts. Along with Torrance, Pannenberg adopts the field concept of the modern physics into his doctrine of creation. By this term, he wants to describe the creative presence and activity of the Spirit in the world. He links the dynamic of the Spirit as a field to time and space. He asserts that the idea of God's transcendence rather demands space. By the omnipresence of God, Pannenberg holds that God must be conceived of as present in each part of space. This explanation of God's omnipresence leads him to a sacramental view of nature. His theological system assumes a color of panentheism. Unlike Pannenberg, Moltmann proclaims himself to be a proponent for panentheism in public. He seems to take the traditional view on creation as a free act of God. But he admits God in some sense needs the world. By the kabalistic term of *zimzum*, he maintains that God makes room for creatures within himself before creation. Since creation, the world is in God. By the term of the cosmic Shekinah, Moltmann claims God will be all in all. His eschatological vision is expressed with the

Conclusion 265

term 'God in the world.' The main goal of his theology is God's immanence.

Despite some suggestive insights, Torrance's position on God's spatiality is too incomplete to speak about it affirmatively. Contrary to him, Pannenberg and Moltmann definitely opt for divine spatiality. But they seem to go too far. They accept the position of panentheism unconsciously or consciously. Around the borderline between Torrance on the one hand and Pannenberg or Moltmann on the other hand, I think, there is a proper answer to this discussion of God's spatiality.

I examined various options and possibilities of God's spatiality. Moltmann's contention that God is space is a mistake that identifies the divine perichoretic space with the triune God himself. We can exist in the divine space but cannot be within God spatially and literally. Through the Spirit, we can dwell in the triune God. Thus, the option that God is space is to be understood metaphorically. Otherwise, this position is to be discarded. Concerning Newton's contention that space is the *sensorium Dei*, there is some controversy over the real intention of Newton. If he understand the absolute space as a kind of God's sense organ, this contention connotes a panentheistic idea. But if his contention implies that God uses space as the medium of his creative presence at the finite places, it is acceptable. In this case, however, we need some other concept of space other than that of the absolute space. I supported the position that God has his own space other than that of creatures. I appealed to the dimensional difference between God and the world. If God has his own space, his space is different from that of creatures. God is not a bodily being. But God is not hostile to body but is related to the material world positively. Because he is the Creator of the world, God is concerned about our bodily life in this world. To occupy a place or space, creatures need a kind of material body. Without a body, however, God can exist in space and interact with the universe. By the term of perichoresis, we designate intra-trinitarian relations in God. The three Persons of the Triune God mutually indwell one another. We also indwell the triune God by the Spirit. Finally, I contended for the view that God is in space of this world. The crucial point is the transcendence of God over the world of space and time. Even though he is present in the limited space of creatures, God should be still transcendent over our space. A dimensional model of relating God and the world represents an

266 *Transcendence and Spatiality of the Triune Creator*

example of the reasonable discourse of God's immanence or presence and transcendence.

William A. Dyrness indicates that our bodily life on the earth is the embodied worship of God. He emphasizes our bodily life by the three categories which are grounded in the triune God's presence in the world: relationship, agency, and embodiment. Spatial and tactual dimensions must be added to our intellectual map and our spiritual journey. Tracy opposes the embodiment of God. But his main target is not against Dyrness's view on embodiment but that of process theologians that the world is the body of God. Even Thomas F. Tracy does affirm some kind of God's embodiment. Deploying fully the three categories of Dyrness, I presented some positive meanings of God's spatiality. To be present in the world, God needs some kind of spatiality. God is not only the creator of the world. But he is also the preserver. While God is transcendent over the world of space and time, he is immanent and present in the world of space and time. Dyrness's categories of relationship, agency, and embodiment are grounded in the presence of the triune God. God is not only intra-trinitarianly relational but also has an internal relationship with the world of creation. God interacts with the world. Even though he cannot be considered as one of the creaturely causes, God can act as an agent. The world is not a body of God. But God embodies his glory through the material things including human body.

I asserted that a kind of divine spatiality is required for the presence of the triune God. In order to be present in the world of space and time, God is spatial or has his own spatiality. According to three categories of Dyrness, I presented three meanings of God's spatiality. The triune God is not an isolated being but a relational being and has a kind of relationality. For this relationship, God needs his own spatiality. By the term of God's spatiality, thus, I emphasized that the triune God exists in a relation or has a relationality. God's relationality is the first meaning of divine spatiality.

The development of modern natural sciences narrowed down the scope of God's action and finally exiled him out of the universe. But the universe as a closed system is now refuted by scientists themselves. This world is not closed but open-ended and the events of this world are not causally determined but indeterministic. Hence there are space, room, or "gaps" for God's action in this world. Developments in such areas as biology, quantum theory and chaos theory are held to signal conceptions

Conclusion

267

of a new openness in the shape of reality and gaps for God's action. The God who is active in this world or the God as an agent has space, room, or gaps for his own activities within the world or universe. This is the second meaning of God's spatiality.

God is not a bodily being. Nor does he have his own body. The world is not to be seen as the body of God. In this world, however, God has embodied himself in various ways. Creation is the theater for God's glory. In the incarnation of the Son, God embodied himself in space and time. The church of both Old Testament and New Testament is the earthly-historical form of God's own existence. The Eucharist is the visual and material means of God's invisible grace. Christ is present in the Eucharist. New creation is the work of making the creation a suitable vehicle for God's glory. It will be an embodied event. In some sense, it has already started. God uses space, time and bodies as vehicles of his grace. By means of them, God expresses his goodness and love for the creature. Thus, we refute the idea of the world as God's body, that is, the literal or direct embodiment of God in the world but acknowledge a kind of God's embodiment. In these processes of embodiment, God uses the material beings including human body as vehicles for his creative presence at the finite space. This is the third and last meaning of divine spatiality.

In the introduction, I stated two purposes of this book. In relation to the book of Ted Peters which presented the relationality and temporality of the divine life, my present study has explored the divine spatiality. By the term of relationality, Peters seems to imply a kind of spatiality. Like the other theologians such as Peacocke, Polkinghorne, and Tracy, however, Peters does not present a divine spatiality boldly. And in relation to the book of Dyrness, which contended that our bodily life is grounded in the presence of the triune God, this study has developed three positive meanings of God's spatiality according to Dyrness's three categories. God uses the limited space of the universe and can be present in it. God is our dwelling place. Thus, God is figuratively a space for us. "The name of the LORD is a strong tower; the righteous run to it and are safe" (Prov. 18:10). There is no 'empty' space, as Einstein states, and there is no space without a field. Through the field theory of the modern physics, the term perichoresis that originally indicates the intra-trinitarian relations is adopted to express a characteristic of the universe of space and time in which we live. By means of the perichoretic concept of space, the co-presence of the infinite and the finite becomes possible.

Finally, I assert that, even though classical theism has some minor flaws, it still provides the effective tool to interpret God's relation to the world. In classical theism, there is a creative tension between God's transcendence and his immanence. I acknowledge the undue emphasis on divine transcendence of traditional or classical theism. In the milieu of pantheism in the pagan world, I think classical theists rightly emphasized the doctrine of transcendence. In comparison to God's transcendence, however, the doctrine of God's immanence was not fully developed in the system of classical theism. Thus, one of the tasks of modern theologians is to fully deploy divine immanence. Panentheism is a kind of overreaction to the above mentioned weak point of classical theism. It emphasizes God's immanence rather than his transcendence and contributes to contemporary theological discussion. Panentheism enables the compensation of the weakness of classical theism. However, panentheism's doctrine of divine transcendence is troublesome. It cannot affirm the transcendence of God distinguished from the world of his creatures. In this sense, I think, panentheists are not fair in handling classical theism. They are not to discard classical theism promptly.

Bibliography

Anderson, Bernhard W., *From Creation to New Creation*, Minneapolis: Fortress Press, 1994.

Anderson, Ray S., *Historical Transcendence and the Reality of God: A Christological Critique*, Grand Rapids: Eerdmans, 1975.

Barbour, Ian G., *Religion and Science: Historical and Contemporary Issues*, San Francisco: HarperCollins, 1997.

Bauckham, Richard, *The Theology of Jürgen Moltmann*, Edinburgh: T&T Clark, 1995.

–, ed., *God Will Be All in All: The Eschatology of Jürgen Moltmann*, Edinburgh: T&T Clark, 1999.

Bertoloni Meli, Domenico, "Caroline, Leibniz, and Clarke," *Journal of the History of Ideas*, 1999, pp. 469–486.

Bloesch, Donald G., *God the Almighty: Power, Wisdom, Holiness, Love*, Downers Grove, IL: InterVarsity Press, 1995.

Boff, Leonardo, *Ecology and Liberation: A New Paradigm (Ecology and Justice)*, trans. John Cumming, Maryknoll, NY: Orbis Books, 1995.

Bonhoeffer, Dietrich, *Discipleship*, trans. Barbara Green and Reinhard Krauss, *Dietrich Bonhoeffer Works*, vol. 4, eds Martin Kuske and Ilse Tödt, Minneapolis: Fortress Press, 2001.

Bouma-Prediger, Steven, *The Greening of Theology: The Ecological Models of Rosemary Radford Ruether, Joshep Sittler, and Jürgen Moltmann*, Atlanta: The American Academy of Religion, 1995.

Bray, Gerald, *The Doctrine of God*, Downers Grove, IL: Inter Varsity Press, 1993.

–, *The Personal God: Is the Classical Understanding of God Tenable?*, Cumbria: Paternoster Press, 1998.

Buller, Cornelius A., *The Unity of Nature and History in Pannenberg's Theology*, Lanham, ML/London: Rowman & Littlefield, 1996.

Burhenn, Herbert, "Pannenberg's doctrine of God," *Scottish Journal of Theology* 28 No 6: pp. 535–549, 1975.

Clayton, Philip, *God and Contemporary Science*, Grand Rapids: Eerdmans, 1997.

270 *Transcendence and Spatiality of the Triune Creator*

Cobb, John B., Jr. and Griffin, David Ray, *Process Theology: An Introductory Exposition*, Philadelphia: The Westminster Press, 1976.

Cobb, John B., Jr. and Pinnock, Clark H., eds, *Searching for an Adequate God: A Dialogue between Process and Free Will Theists*, Grand Rapids: Eerdmans, 2000.

Coffey, David, "The 'Incarnation' of the Holy Spirit in Christ," *Theological Studies* 45, 1984, pp. 466–480.

Coles, Peter, *Hawking and the Mind of God*, New York: Totem Books, 2000.

Colyer, Elmer M., *How to Read T. F. Torrance: Understanding His Trinitarian and Scientific Theology*, Downers Grove, IL: InterVarsity Press, 2001.

Cooper, John W., *Our Father in Heaven: Christian Faith and Inclusive Language for God*, Grand Rapids: Baker Books, 1998.

Davies, Paul, *Space and Time in the Modern Universe*, Cambridge University Press, 1977.

–, *God and the New Physics*, New York: Simon & Schuster, Inc., 1984.

Davis, Edward B., "Newton's Rejection of the 'Newtonian World View': The Role of Divine Will in Newton's Natural Philosophy," *Fides et Historia* 22, Summer 1990, pp. 6–20.

Dyrness, William A., *The Earth is God's: A Theology of American Culture*, Maryknoll, NY: Orbis Books, 1997.

–, *Visual Faith: Art, Theology, and Worship in Dialogue*, Grand Rapids: Baker Books, 2001.

Eliade, Mircea, *The Sacred and the Profane: The Nature of Religion*, trans. Williard R. Trask, New York: Harourt, Brace, 1959.

Einstein, Albert, *Relativity: The Special and the General Theory*, trans. Robert W. Lawson, New York: Three Rivers Press, 1961.

–, *The World as I See It*, trans. Alan Harris, New York: Kensington, 2000.

Farrow, Douglas, "Between the Rock and a Hard Place: In Support of (something like) a Reformed View of the Eucharist," *International Journal of Systematic Theology* 3:2 (July 2001).

Ford, David F., *The Modern Theologians: An Introduction to Christian Theology in the Twentieth Century*, Oxford: Blackwell, 1997, Second Edition.

Gardner, Sebastian, *Kant and the Critique of Pure Reason*, London and New York: Routledge, 1999.

Bibliography

Goldberg, Michael, "God, Action and Narrative: *Which* Narrative? *Which* Action? *Which* God?" *The Journal of Religion* 68, January 1988, pp. 39–56.

Grenz, Stanley J., *Reason for Hope: the Systematic Theology of Wolfhart Pannenberg*, New York: Oxford University Press, 1990.

–, *Theology for the Community of God*, Grand Rapids/Vancouver: Eerdmans/Regent College Publishing, 1994.

Grenz, Stanley J. and Olson, Roger E., *20th-Century Theology: God and the World in a Transitional Age*, Downers Grove: InterVarsity Press, 1992.

Gunton, Colin E., ed., *The Cambridge Companion to Christian Doctrine*, Cambridge University Press, 1997.

–, *Becoming and Being: The Doctrine of God in Charles Hartshorne and Karl Barth*, Oxford: Oxford University Press, 1978.

–, *Yesterday and Today: A Study of Continuities in Christology*, Grand Rapids: Eerdmans, 1983.

–, *The Actuality of Atonement: A Study of Metaphor, Rationality and the Christian Tradition*, Grand Rapids: Eerdmans, 1989.

–, "Relation and Relativity: The Trinity and the Created World," Christoph Schwöbel ed., *Trinitarian Theology Today: Essays on Divine Being and Act*, Edinburgh: T&T Clark, 1995.

–, *The Triune Creator: A Historical and Systematic Study*, Grand Rapids: Eerdmans, 1998.

–, ed., *Trinity, Time, and Church: A Response to the Theology of Robert W. Jenson*, Grand Rapids/Cambridge: Eerdmans, 2000

–, *Christ and Creation*, Grand Rapids: Eerdmans/Carlisle: Paternoster, 1992.

Guthrie, Donald, *New Testament Theology*, Downers Grove: InterVarsity Press, 1981.

Helm, Paul, *Calvin and the Calvinists,* Carlisle, PA/Edinburgh: The Banner of Truth Trust, 1982.

–, *Eternal God: A Study of God without Time*, Oxford University Press, 1988.

–, *The Providence of God*, Downers Grove, IL: InterVarsity Press, 1994.

Jammer, Max, *Concepts of Space: The History of Theories of Space in Physics*, New York: Dover Publications, Inc., 1993, 3rd, enlarged Edition.

272 *Transcendence and Spatiality of the Triune Creator*

Jantzen, Grace, *God's World, God's Body*, Philadelphia: Westminster Press, 1984.

Jenson, Robert W., "The Body of God's Presence: A Trinitarian Theory," *Creation, Christ, and Culture: Studies in Honour of T. F. Torrance*, ed., Richard W. A. McKinney, Edinburgh: T&T Clark, 1976.

Johnson, Elizabeth A., *SHE WHO IS: The Mystery of God in Feminist Theological Discourse*, New York: Crossroad, 1992.

Johnson, Mark, *The Body in the Mind: The bodily Basis of Meaning, Imagination, and Reason*, Chicago & London: The University of Chicago Press, 1987.

Kant, Immanuel, *The Critique of Pure Reason*, trans. Norman K. Smith, New York: The Humanities Press, 1950.

Kasper, Walter, *The God of Jesus Christ*, New York: Crossroad, 1986.

Kaufman, Gordon D., *God the Problem*, Cambridge, MA: Harvard University Press, 1972.

Kimel, Alvin F. Jr., ed., *Speaking the Christian God: The Holy Trinity and the Challenge of Feminism*, Grand Rapids: Eerdmans, 1992.

Kreeft, Peter and Tacelli, Ronald K., *Handbook of Christian Apologetics: Hundreds of Answers to Crucial Questions*, Downers Grove: InterVarsity Press, 1994.

LaCugna, Catherine Mowry, *God for Us: The Trinity and Christian Life*, New York: HarperCollins Publishers, 1991.

Ladd, George Eldon, *A Theology of the New Testament Theology*, Grand Rapids: Eerdmans, Revised Edition, ed., Donald A. Hagner, 1994.

Lancaster, Sarah Heaner, "Divine Relations of the Trinity: Augustine's Answer to Arianism," *Calvin Theological Journal* 34:2 (1999), pp. 327–346.

McFague, Sallie, *Metaphorical Theology: Models of God in Religious Language*, Philadelphia: Fortress Press, 1982.

–, *Models of God: Theology for an Ecological, Nuclear Age*, Philadelphia: Fortress Press, 1987.

–, *The Body of God: An Ecological Theology*, Minneapolis: Fortress Press, 1993.

McGrath, Alister E., *T. F. Torrance: An Intellectual Biography*, Edinburgh: T&T Clark, 1999.

Moltmann, Jürgen, *The Church in the Power of the Spirit: A Contribution to Messianic Ecclesiology*, trans. M. Kohl, Minneapolis: Fortress Press, 1993 (In German, 1975).

Bibliography 273

–, *The Trinity and the Kingdom: The Doctrine of God*, trans. M. Kohl, Minneapolis: Fortress Press, 1993 (In German, 1980).

–, *History and the Triune God: Contributions to Trinitarian Theology*, trans. John Bowden, New York: Crossroad, 1992.

–, *God in Creation*, trans. M. Kohl, Minneapolis: Fortress Press, 1993 (In German, 1985).

–, *The Coming of God*, trans. M. Kohl, Minneapolis: Fortress Press, 1996 (In German, 1995).

–, *Experiences in Theology*, trans. M. Kohl, Minneapolis: Fortress Press, 2000.

Morris, Thomas V., *Our Idea of God: An Introduction to Philosophical Theology*, Downers Grove, IL: InterVarsity Press, 1991.

Muller, Richard, *Dictionary of Latin and Greek Theological Terms: Drawn Principally from Protestant Scholastic Theology*, Grand Rapids: Baker Books, 1985.

–, "Incarnation, Immutability, and the Case for Classical Theism," *Westminster Theological Journal* 45, 1983.

Murphy, Nancey, *Beyond Liberalism and Fundamentalism: How Modern and Postmodern Philosophy Set the Theological Agenda*, Valley Forge, PA: Trinity Press International, 1996.

–, *Reconciling Theology and Science: A Radical Reformation Perspective*, Kitchener: Pandora Press, 1997.

–, *Anglo-American Postmodernity: Philosophical Perspectives on Science, Religion, and Ethics*, Boulder: Westview Press, 1997.

–, "Divine Action in the Natural Order: Buridan's Ass and Schrödinger's Cat," R. J. Russell, N. Murphy, and A. R. Peacocke, eds, *Chaos and Complexity: Scientific Perspectives on Divine Action*, Vatican City State: Vatican Observatory Publications, 1997, 2nd Ed.

Murphy, Nancey and Ellis, George F. R., *On the Moral Nature of the Universe: Theology, Cosmology, and Ethics*, Minneapolis: Fortress Press, 1996.

Nash, Ronald H., *The Concept of God: An Exploration of Contemporary Difficulties with the Attributes of God*, Grand Rapids: Zondervan, 1983.

–, ed., *Process Theology*, Grand Rapids: Baker Books, 1987.

Neville, Robert C., *God the Creator: On the Transcendence and Presence of God*, Chicago & London: The University of Chicago Press, 1968.

274 *Transcendence and Spatiality of the Triune Creator*

O'Donnell, John J., *Trinity and Temporality: The Christian Doctrine of God in the Light of Process Theology and the Theology of Hope*, Oxford University Press, 1983.

Olson, Roger E., "Trinity and Eschatology: the Historical Being of God in Jürgen Moltmann and Wolfhart Pannenberg," *Scottish Journal of Theology* 36 No 2: pp. 213–227, 1983.

–, "Trinity and Eschatology: The Historical Being of God in the Theology of Wolfhart Pannenberg," Rice University Ph. D. dissertation, 1984.

–, "The Human Self-Realization of God: Hegelian Elements in Pannenberg's Christology," *Perspectives in Religious Studies* 13: pp. 207–223, 1986.

–, "Wolfhart Pannenberg's Doctrine of the Trinity," *Scottish Journal of Theology* 43 No 2: pp. 175–206, 1990.

Pannenberg, Wolfhart, *Toward a Theology of Nature: Essays on Science and Faith*, ed. T. Peters, Louisville, KY: Westminster/John Knox Press, 1993.

–, *Systematic Theology* vol. 1, trans. G. W. Bromiley, Grand Rapids: Eerdmans, 1991 (German, 1988).

–, *Systematic Theology* vol. 2, trans. G. W. Bromiley, Grand Rapids: Eerdmans, 1994 (German, 1991).

–, *Systematic Theology* vol. 3, trans. G. W. Bromiley, Grand Rapids: Eerdmans, 1998 (German, 1993).

–, "The Doctrine of Creation and Modern Science," Ted Peters, ed., *Cosmos as Creation*, Nashiville: Abingdon Press, 1989.

–, "Theology and Science," *Princeton Seminary Bulletin* 13 No 3, 1992.

Peacocke, Arthur, *Intimations of Reality: Critical Realism in Science and Religion*, Notre Dame: University of Notre Dame Press, 1984.

–, *Theology for a Scientific Age: Being and Becoming – Natural, Divine, and Human*, Minneapolis: Fortress Press, 1993, enlarged Ed.

–, "God's Interaction with the World: The Implications of Deterministic 'Chaos' and of Interconnected and Interdependent Complexity," R. J. Russell, N. Murphy, and A. R. Peacocke, eds, *Chaos and Complexity: Scientific Perspectives on Divine Action*, Vatican City State: Vatican Observatory Publications, 1997, 2nd Ed.

Peters, Ted, *God as Trinity: Relationality and Temporality in Divine Life*, Louisville, KN: Westminster/John Knox Press, 1993.

Bibliography

Pinnock, Clark, et al., *The Openness of God: A Biblical Challenge to the Traditional Understanding of God*, Downers Grove, IL: IVP, 1994.

Polkinghorne, John, *Science and Creation: The Search for Understanding*, London: SPCK, 1988.

–, *Science and Providence: God's Interaction with the World*, London: SPCK, 1989.

–, *Reason and Reality: The Relationship between Science and Theology*, Valley Forge, PA: Trinity Press International, 1991.

–, *Belief in God in an Age of Science*, New Haven & London: Yale University Press, 1998.

–, ed., *The Work of Love: Creation as Kenosis*, Grand Rapids/ Cambridge: Eerdmans/SPCK, 2001.

Preuss, Horst Dietrich, *Old Testament Theology*, vol. 1, Louisville, KY: Westminster John Knox Press, 1995, trans. Leo G. Perdue.

Purves, Andrew, "The Christology of Thomas F. Torrance," ed. Elmer Colyer, *The Promise of Trinitarian Theology: Theologians in Dialogue with Thomas F. Torrance*, Lanham, MD: Rowman& Littlefield, 2001.

Quine, W. V. O., "Two Dogmas of Empiricism," *From a Logical Point of View: Logico-Philosophical Essays*, Cambridge, MA: Harvard University Press, 1953.

Russell, R. J., N. Murphy & A. R. Peacocke, eds, *Chaos and Complexity: Scientific Perspectives on Divine Action*, Vatican City State: Vatican Observatory Publications/Berkeley, CA: The Center for Theology and the Natural Sciences, 1997, 2nd Ed.

Schawarz, Hans, "God's Place in a Space Age," *Zygon* 21:3, September 1986, pp. 353–368.

Schwöbel, Christoph, ed., *Trinitarian Theology Today: essays on Divine Being and Act*, Edinburgh: T&T Clark, 1995.

Thiel, John E., *God and World in Schleiermacher's Dialektik and Glaubenslehre: Criticism and the Methodology of Dogmatics*, Peter Lang, 1981.

Thomas, Owen C., "Chaos, Complexity, and God: A Review Essay," *Theology Today* 54, April 1997, pp. 66–76.

Torrance, Thomas, F., *Space, Time and Incarnation*, Edinburgh: T&T Clark, 1997.

–, *Space, Time and Resurrection*, Edinburgh: T&T Clark, 1998.

–, *Divine and Contingent Order*, Edinburgh: T&T Clark, 1998.

276   *Transcendence and Spatiality of the Triune Creator*

–, *The Trinitarian Faith: The Evangelical Theology of Ancient Catholic Church*, Edinburgh: T&T Clark, 1995.

–, *The Christian Doctrine of God: One Being Three Persons*, Edinburgh & New York: T&T Clark, 2001.

–, *Reality and Evangelical Theology: The Realism of Christian Revelation*, Downers Grove, IL: InterVarsity Press, 1999.

Tracy, Thomas F., *God, Action, and Embodiment*, Grand Rapids: Eerdmans, 1984.

–, ed., *The God who Acts: Philosophical and Theological Explorations*, Penn State University Press, 1994.

–, "Particular Providence and the God of the Gaps," R. J. Russell, N. Murphy, and A. R. Peacocke, eds, *Chaos and Complexity: Scientific Perspectives on Divine Action*, Vatican City State: Vatican Observatory Publications, 1997, 2nd Ed.

Viladesau, Richard, *Theological Aesthetics: God in Imagination, Beauty, and Art*, New York: Oxford University Press, 1999.

Volf, Miroslav, *After Our Likeness: The Church as the Image of the Trinity*, Grand Rapids: Eerdmans, 1998.

Von Rad, Gerhard, *Old Testament Theology*, vol. 1, trans., D. M. G. Stalker, San Francisco: HarperCollins, 1962.

–, *Old Testament Theology*, vol. 2, trans., D. M. G. Stalker, San Francisco: HarperCollins, 1965.

Wainwright, Geoffrey, *Eucharist and Eschatology*, New York: Oxford University Press, 1981.

Wainwright, William J., "God's Body," *The Concept of God*, ed. T. Morris, Oxford: Oxford University Press, 1987.

Welker, Michael, *God the Spirit*, trans. John F. Hoffmeyer, Minneapolis: Fortress Press, 1994.

Williams, Robert R., *Schleiermacher the Theologian: The Construction of the Doctrine of God*, Philadelphia: Fortress Press, 1978.

Worthing, Mark William, *God, Creation, and Contemporary Physics*, Minneapolis, MN: Augsburg Fortress, 1996.